Introduction to Topology

Third Edition

by Bert Mendelson

DOVER PUBLICATIONS, INC.
NEW YORK

This Dover edition, first published in 1990, is an unabridged, unaltered republication of the work originally published by Allyn and Bacon, Inc., Boston, 1975 (First Edition, 1962; Second Edition, 1968).

Manufactured in the United States of America
Dover Publications, Inc., 31 East 2nd Street, Mineola, N.Y. 11501

Library of Congress Cataloging-in-Publication Data

Mendelson, Bert, 1926–1988
Introduction to topology / by Bert Mendelson. — 3rd ed.
p. cm.
"Unabridged, unaltered republication of the work originally published by Allyn and Bacon, Inc., Boston, 1975"—T.p. verso.
Includes bibliographical references (p.).
ISBN 0-486-66352-3
1. Topology. I. Title.
[QA611.M39 1990]
514—dc20 90-32980
CIP

Contents

Preface *ix*

1 Theory of Sets 1

1 Introduction 1
2 Sets and subsets 2
3 Set operations: union, intersection, and complement 4
4 Indexed families of sets 7
5 Products of sets 9
6 Functions 11
7 Relations 15
8 Composition of functions and diagrams 17
9 Inverse functions, extensions, and restrictions 21
10 Arbitrary products 25

2 Metric Spaces 29

1 Introduction 29

2 Metric spaces 30
3 Continuity 35
4 Open balls and neighborhoods 40
5 Limits 47
6 Open sets and closed sets 52
7 Subspaces and equivalence of metric spaces 58
8 An infinite dimensional Euclidean space 66

3 Topological Spaces 70

1 Introduction 70
2 Topological spaces 71
3 Neighborhoods and neighborhood spaces 75
4 Closure, interior, boundary 81
5 Functions, continuity, homeomorphism 87
6 Subspaces 92
7 Products 97
8 Identification topologies 101
9 Categories and functors 107

4 Connectedness 112

1 Introduction 112
2 Connectedness 113
3 Connectedness on the real line 119
4 Some applications of connectedness 122
5 Components and local connectedness 130
6 Path-connected topological spaces 133
7 Homotopic paths and the fundamental group 139
8 Simple connectedness 151

5 Compactness 157

1 Introduction 157
2 Compact topological spaces 158

3 Compact subsets of the real line 164
4 Products of compact spaces 168
5 Compact metric spaces 172
6 Compactness and the Bolzano-Weierstrass property 179
7 Surfaces by identification 186

Bibliography 201

Index 203

Preface to the Third Edition

The first edition of this text was based on lecture notes prepared for a one-semester undergraduate course given at Smith College. The aim was to present a simple, thorough survey of elementary topics to students whose preparation included a calculus sequence in which some attention had been paid to definitions and proofs of theorems. With this in mind, I have attempted to resist the temptation to include more topics. There are many excellent introductory topology texts which are first-year graduate school level texts and it was not my original intention, nor is it now, to write at that level.

The main outlines of the text have not been changed. The first chapter is an informal discussion of set theory. The concept of countability has been postponed until Chapter 5, where it appears in the context of compactness.

The second chapter is a discussion of metric spaces. The topological terms "open set," "neighborhood," etc., are introduced. Particular attention is paid to various distance functions which may be defined on Euclidean n-space and which lead to the ordinary topology.

In taking up topological spaces in Chapter 3, the transition from the particular to the general has been maintained, so that the concept of topological space is viewed as a generalization of the concept of metric space. Thus there is a similarity or, perhaps, a redundancy in the presentation of these two topics. A great deal of attention has been paid to alternate procedures for the creation of a topological space, using neighborhoods, etc., in the hope that this seemingly trivial, but subtle, point would be clarified.

Chapters 4 and 5 are devoted to a discussion of the two most important topological properties: connectedness and compactness. Some of this material could lead to further discussion of topics related to analysis, function spaces, separation axioms, metrization theorems, to name a few. On the other hand, material such as homotopy and two-dimensional closed surfaces could lead to further discussion of topics related to algebraic topology.

In conclusion, it is a pleasure to express in print my gratitude to those mathematicians under whom I studied and who helped make this book possible. In particular I should like to mention Professors C. Chevalley, S. Eilenberg, I. James, H. Riberio, P. Smith, and E. Thomas.

B. M.

Introduction to Topology

CHAPTER 1

Theory of Sets

1 INTRODUCTION

As in any other branch of mathematics today, topology consists of the study of collections of objects that possess a mathematical structure. This remark should not be construed as an attempt to define *mathematics*, especially since the phrase "mathematical structure" is itself a vague term. We may, however, illustrate this point by an example.

The set of *positive integers* or *natural numbers* is a collection of objects N on which there is defined a function s, called the *successor function*, satisfying the conditions:

1. For each object x in N, there is one and only one object y in N such that $y = s(x)$;
2. Given objects x and y in S such that $s(x) = s(y)$, then $x = y$;
3. There is one and only one object in N, denoted by 1, which is not the successor of an object in N, i.e., $1 \neq s(x)$ for each x in N;
4. Given a collection T of objects in N such that 1 is in T and for each x in T, $s(x)$ is also in T, then $T = N$.

The four conditions enumerated above are referred to as *Peano's axioms for the natural numbers*. The fourth condition is called *the principle of mathematical induction*. One defines addition of natural numbers in such a manner that $s(x) = x + 1$, for each x in N, which explains the use of the word "successor" for the function s. What is significant at the moment is the conception of the natural numbers as constituting a certain collection of objects N with an additional mathematical structure, namely the function s.

We shall describe a *topological space* in the same terms, that is, a collection of objects together with a specified structure. A topological space is a collection of objects (these objects usually being referred to as points), and a structure that endows this collection of points with some coherence, in the sense that we may speak of nearby points or points that in some sense are close together. This structure can be prescribed by means of a collection of subcollections of points called *open sets*. As we shall see, the major use of the concept of a topological space is that it provides us with an exact, yet exceedingly general setting for discussions that involve the concept of continuity.

By now the point should have been made that topology, as well as other branches of mathematics, is concerned with the study of collections of objects with certain prescribed structures. We therefore begin the study of topology by first studying collections of objects, or, as we shall call them, *sets*.

2 SETS AND SUBSETS

We shall assume that the terms "object," "set," and the relation "is a member of" are familiar concepts. We shall be concerned with using these concepts in a manner that is in agreement with the ordinary usage of these terms.

If an object A belongs to a set S we shall write $A \in S$ (read, "A in S"). If an object A does not belong to a set S we shall write $A \notin S$ (read, "A not in S"). If $A_1, \ldots, A_n$ are objects, the set consisting of precisely these objects will be written $\{A_1, \ldots, A_n\}$. For purposes of logical precision it is often necessary to distinguish the set $\{A\}$, consisting of precisely one object A, from the object A itself. Thus $A \in \{A\}$ is a true statement, whereas $A = \{A\}$ is a false statement. It is also necessary that there be a set that has no members, the so-called *null* or *empty* set. The symbol for this set is $\emptyset$.

Let A and B be sets. If for each object $x \in A$, it is true that $x \in B$, we say that A is a *subset* of B. In this event, we shall also say that A *is contained in* B, which we write $A \subset B$, or that B *contains* A, which we write $B \supset A$.

In accordance with the definition of subset, a set A is always a subset of itself. It is also true that the empty set is a subset of A. These two subsets, A and $\emptyset$, of A are called *improper* subsets, whereas any other subset is called a *proper* subset.

There are certain subsets of the real numbers that are frequently considered in calculus. For each pair of real numbers a, b with $a < b$, the set of all real numbers x such that $a \leqq x \leqq b$ is called the *closed interval* from a to b and is denoted by $[a, b]$. Similarly, the set of all real numbers x such that $a < x < b$ is called the *open interval* from a to b and is denoted by (a, b). We thus have $(a, b) \subset [a, b] \subset R$, where R is the set of real numbers.

Two sets are identical if they have precisely the same members. Thus, if A and B are sets, $A = B$ if and only if both $A \subset B$ and $B \subset A$. Frequent use is made of this fact in proving the equality of two sets.

Sets may themselves be objects belonging to other sets. For example, $\{\{1, 3, 5, 7\}, \{2, 4, 6\}\}$ is a set to which there belong two objects, these two objects being the set of odd positive integers less than 8 and the set of even positive integers less than 8. If A is any set, there is available as objects with which to constitute a new set, the collection of subsets of A. In particular, for each set A, there is a set we denote by 2^A whose members are the subsets of A. Thus, for each set A, we have $B \in 2^A$ if and only if $B \subset A$.

EXERCISES

1. Determine whether each of the following statements is true or false:
   (a) For each set A, $A \in 2^A$.
   (b) For each set A, $A \subset 2^A$.
   (c) For each set A, $\{A\} \subset 2^A$.
   (d) For each set A, $\emptyset \in 2^A$.
   (e) For each set A, $\emptyset \subset 2^A$.
   (f) There are no members of the set $\{\emptyset\}$.
   (g) Let A and B be sets. If $A \subset B$, then $2^A \subset 2^B$.
   (h) There are two distinct objects that belong to the set $\{\emptyset, \{\emptyset\}\}$.
2. Let A, B, C be sets. Prove that if $A \subset B$ and $B \subset C$, then $A \subset C$.
3. Let $A_1, \ldots, A_n$ be sets. Prove that if $A_1 \subset A_2$, $A_2 \subset A_3, \ldots,$ $A_{n-1} \subset A_n$ and $A_n \subset A_1$, then $A_1 = A_2 = \ldots = A_n$.

3 SET OPERATIONS: UNION, INTERSECTION, AND COMPLEMENT

If x is an object, A a set, and $x \in A$, we shall say that x is an *element*, *member*, or *point* of A. Let A and B be sets. The *intersection* of the sets A and B is the set whose members are those objects x such that $x \in A$ and $x \in B$. The intersection of A and B is denoted by $A \cap B$ (read, "A intersect B"). The *union* of the sets A and B is the set whose members are those objects x such that x belongs to at least one of the two sets A, B; that is, either $x \in A$ or $x \in B$.* The union of A and B is denoted by $A \cup B$ (read, "A union B").

The operations of set union and set intersection may be rep-

* The logical connective "or" is used in mathematics (and also in logic) in the inclusive sense. Thus, a compound statement "P or Q" is true in each of the three cases: (1) P true, Q false; (2) P false, Q true; (3) P true, Q true, whereas "P or Q" is false only if both P and Q are false.

resented pictorially (by *Venn diagrams*). In Figure 1, let the elements of the set A be the points in the region shaded by lines running from the lower left-hand part of the page to the upper right-hand part of the page, and let the elements of the set B be the points in the region shaded by lines sloping in the opposite direction. Then the elements of $A \cup B$ are the points in either shaded region and the elements of $A \cap B$ are the points in the cross-hatched region.

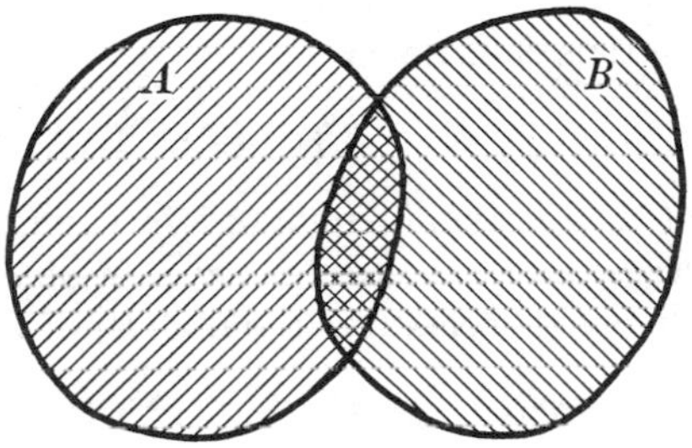

Figure 1

Let $A \subset S$. The *complement* of A in S is the set of elements that belong to S but not to A. The complement of A in S is denoted by $C_S(A)$, S/A, or by $S - A$. The set S may be fixed throughout a given discussion, in which case the complement of A in S may simply be called the complement of A and be denoted by $C(A)$. $C(A)$ is again a subset of S and one may take its complement. The complement of the complement of A is A; that is, $C(C(A)) = A$.

There are many formulas relating the set operations of intersection, union, and complementation. Frequent use is made of the following two formulas.

THEOREM (DeMorgan's Laws). Let $A \subset S$, $B \subset S$. Then

$$C(A \cup B) = C(A) \cap C(B), \tag{3.1}$$

$$C(A \cap B) = C(A) \cup C(B). \tag{3.2}$$

Proof. Suppose $x \in C(A \cup B)$. Then $x \in S$ and $x \notin A \cup B$. Thus, $x \notin A$ and $x \notin B$, or $x \in C(A)$ and $x \in C(B)$. Therefore $x \in C(A) \cap C(B)$ and, consequently,

$$C(A \cup B) \subset C(A) \cap C(B).$$

Conversely, suppose $x \in C(A) \cap C(B)$. Then $x \in S$ and $x \in C(A)$ and $x \in C(B)$. Thus, $x \notin A$ and $x \notin B$, and therefore $x \notin A \cup B$. It follows that $x \in C(A \cup B)$ and, consequently,

$$C(A) \cap C(B) \subset C(A \cup B).$$

We have thus shown that

$$C(A) \cap C(B) = C(A \cup B).$$

One may prove Formula 3.2 in much the same manner as 3.1 was proved. A shorter proof is obtained if we apply 3.1 to the two subsets $C(A)$ and $C(B)$ of S, thus

$$C(C(A) \cup C(B)) = C(C(A)) \cap C(C(B)) = A \cap B.$$

Taking complements again, we have

$$C(A) \cup C(B) = C(C(C(A) \cup C(B))) = C(A \cap B).$$

EXERCISES

1. Let $A \subset S$, $B \subset S$. Prove the following:
   (a) $A \subset B$ if and only if $A \cup B = B$.
   (b) $A \subset B$ if and only if $A \cap B = A$.
   (c) $A \subset C(B)$ if and only if $A \cap B = \emptyset$.
   (d) $C(A) \subset B$ if and only if $A \cup B = S$.
   (e) $A \subset B$ if and only if $C(B) \subset C(A)$.
   (f) $A \subset C(B)$ if and only if $B \subset C(A)$.
2. Let $X \subset Y \subset Z$. Prove the following:
   (a) $C_Y(X) \subset C_Z(X)$.
   (b) $Z - (Y - X) = X \cup (Z - Y)$.

4 INDEXED FAMILIES OF SETS

Let I be a set. For each $\alpha \in I$, let A_α be a subset of a given set S. We call I an indexing set and the collection of subsets of S indexed by the elements of I is called an *indexed family* of subsets of S. We denote this indexed family of subsets of S by $\{A_\alpha\}_{\alpha \subset I}$ (read, "A sub-alpha, alpha in I").

Let $\{A_\alpha\}_{\alpha \in I}$ be an indexed family of subsets of a set S. The union of this indexed family, written, $\bigcup_{\alpha \in I} A_\alpha$, (read "union over α in I of A_α") is the set of all elements $x \in S$ such that $x \in A_\beta$ for at least one index $\beta \in I$. The intersection of this indexed family, written $\bigcap_{\alpha \in I} A_\alpha$ (read "intersection over α in I of A_α") is the set of all elements $x \in S$ such that $x \in A_\beta$ for each $\beta \in I$. [Note that $\bigcup_{\alpha \in I} A_\alpha = \bigcup_{\gamma \in I} A_\gamma$, for which reason the two occurrences of "α" in the expression $\bigcup_{\alpha \in I} A_\alpha$ are referred to as dummy indices.]

As an example, let A_1, A_2, A_3, A_4 be respectively the set of freshmen, sophomores, juniors, and seniors in some specified college. Here we have $I = \{1, 2, 3, 4\}$ as an indexing set, and $\bigcup_{\alpha \in I} A_\alpha$ is the set of undergraduates while $\bigcap_{\alpha \in I} A_\alpha = \emptyset$.

If the indexing set I contains precisely two distinct indices, then the union over α in I of A_α is the same as the union of two sets as defined in the previous section; that is,

$$\bigcup_{\alpha \in \{i,j\}} A_\alpha = A_i \cup A_j.$$

Similarly,

$$\bigcap_{\alpha \in \{i,j\}} A_\alpha = A_i \cap A_j.$$

We have allowed for the possibility that the indexing set I is the empty set in which case a careful reading of the definition shows that

$$\bigcup_{\alpha \in \emptyset} A_\alpha = \emptyset.$$

$$\bigcap_{\alpha \in \emptyset} A_\alpha = S.$$

DeMorgan's laws are applicable to unions and intersections of indexed families of subsets of a set S.

THEOREM Let $\{A_\alpha\}_{\alpha \in I}$ be an indexed family of subsets of a set S. Then

(4.1) $$C(\bigcup_{\alpha \in I} A_\alpha) = \bigcap_{\alpha \in I} C(A_\alpha),$$

(4.2) $$C(\bigcap_{\alpha \in I} A_\alpha) = \bigcup_{\alpha \in I} C(A_\alpha).$$

Proof. Suppose $x \in C(\bigcup_{\alpha \in I} A_\alpha)$. Then $x \notin \bigcup_{\alpha \in I} A_\alpha$; that is, $x \notin A_\beta$ for each index $\beta \in I$. Thus $x \in C(A_\beta)$ for each index $\beta \in I$ and $x \in \bigcap_{\alpha \in I} C(A_\alpha)$. Therefore,

$$C(\bigcup_{\alpha \in I} A_\alpha) \subset \bigcap_{\alpha \in I} C(A_\alpha).$$

Conversely, suppose that $x \in \bigcap_{\alpha \in I} C(A_\alpha)$. Then $x \in C(A_\beta)$ for each index $\beta \in I$. Thus $x \notin A_\beta$ for each index $\beta \in I$; that is, $x \notin \bigcup_{\alpha \in I} A_\alpha$. Therefore, $x \in C(\bigcup_{\alpha \in I} A_\alpha)$ and

$$\bigcap_{\alpha \in I} C(A_\alpha) \subset C(\bigcup_{\alpha \in I} A_\alpha).$$

This proves 4.1. The proof of 4.2 is left as an exercise.

Given any collection of subsets of a set S, the concept of indexed family of subsets allows us to define the union or intersection of the aforementioned subsets. We need only construct some convenient indexing set. In the event that the collection of subsets is finite, the finite set $\{1, 2, \ldots, n\}$ of integers is a convenient indexing set. Given subsets $A_1, A_2, \ldots, A_n$ of S, we shall often write $A_1 \cup A_2 \cup \ldots \cup A_n$ or $\bigcup_{i=1}^{n} A_i$ for $\bigcup_{\alpha \in \{1,2,\ldots,n\}} A_\alpha$ and, similarly, $A_1 \cap A_2 \cap \ldots \cap A_n$ or $\bigcap_{i=1}^{n} A_i$ for $\bigcap_{\alpha \in \{1,2,\ldots,n\}} A_\alpha$.

EXERCISES

1. Let $\{A_\alpha\}_{\alpha \in I}$, $\{B_\alpha\}_{\alpha \in I}$ be two indexed families of subsets of a set S. Prove the following:
   (a) For each $\beta \in I$, $A_\beta \subset \bigcup_{\alpha \in I} A_\alpha$.
   (b) For each $\beta \in I$, $\bigcap_{\alpha \in I} A_\alpha \subset A_\beta$.
   (c) $\bigcup_{\alpha \in I} (A_\alpha \cup B_\alpha) = (\bigcup_{\alpha \in I} A_\alpha) \cup (\bigcup_{\alpha \in I} B_\alpha)$.
   (d) $\bigcap_{\alpha \in I} (A_\alpha \cap B_\alpha) = (\bigcap_{\alpha \in I} A_\alpha) \cap (\bigcap_{\alpha \in I} B_\alpha)$.

(e) If for each $\beta \in I$, $A_\beta \subset B_\beta$ then

$$\bigcup_{\alpha \in I} A_\alpha \subset \bigcup_{\alpha \in I} B_\alpha,$$

$$\bigcap_{\alpha \in I} A_\alpha \subset \bigcap_{\alpha \in I} B_\alpha.$$

(f) Let $D \subset S$. Then

$$\bigcup_{\alpha \in I} (A_\alpha \cap D) = \left(\bigcup_{\alpha \in I} A_\alpha\right) \cap D,$$

$$\bigcap_{\alpha \in I} (A_\alpha \cup D) = \left(\bigcap_{\alpha \in I} A_\alpha\right) \cup D.$$

2. Let A, B, $D \subset S$. Then

$$A \cap (B \cup D) = (A \cap B) \cup (A \cap D),$$

$$A \cup (B \cap D) = (A \cup B) \cap (A \cup D).$$

3. Let $\{A_\alpha\}_{\alpha \in I}$ be an indexed family of subsets of a set S. Let $J \subset I$. Prove that
(a) $\bigcap_{\alpha \in J} A_\alpha \supset \bigcap_{\alpha \in I} A_\alpha$.
(b) $\bigcup_{\alpha \in J} A_\alpha \subset \bigcup_{\alpha \in I} A_\alpha$.

4. Let $\{A_\alpha\}_{\alpha \in I}$ be an indexed family of subsets of a set S. Let $B \subset S$. Prove that
(a) $B \subset \bigcap_{\alpha \in I} A_\alpha$ if and only if for each $\beta \in I$, $B \subset A_\beta$.
(b) $\bigcup_{\alpha \in I} A_\alpha \subset B$ if and only if for each $\beta \in I$, $A_\beta \subset B$.

5. Let I be the set of real numbers that are greater than 0. For each $x \in I$, let A_x be the open interval $(0, x)$. Prove that $\bigcap_{x \in I} A_x = \emptyset$, $\bigcup_{x \in I} A_x = I$. For each $x \in I$, let B_x be the closed interval $[0, x]$. Prove that $\bigcap_{x \in I} B_x = \{0\}$, $\bigcup_{x \in I} B_x = I \cup \{0\}$.

5 PRODUCTS OF SETS

Let x and y be objects. The *ordered pair* (x, y)* is a sequence of two objects, the first object of the sequence being x and the second object of the sequence being y. Let A and B be sets. The *Cartesian product* of A and B, written $A \times B$, (read "A cross B") is the set

---

* If x and y are real numbers, the symbol (x, y) is ambiguous, for it may stand for either the ordered pair whose first element is x and the second y, or for the open interval (x, y). It is hoped that this ambiguity will be resolved by the context in which the symbol occurs.

whose elements are all the ordered pairs (x, y) such that $x \in A$ and $y \in B$.

The Cartesian product of two sets is a familiar notion. The coordinate plane of analytic geometry is the Cartesian product of two lines. The possible outcomes of the throw of a pair of dice is the Cartesian product of two sets, A and B, where $A = B = \{1, 2, 3, 4, 5, 6\}$. Unless $A = B$, the two Cartesian products $A \times B$ and $B \times A$ are distinct.

A generalization of the Cartesian product of two sets is the direct product of a sequence of sets. Let $A_1, A_2, \ldots, A_n$ be a finite sequence of sets, indexed by $\{1, 2, \ldots, n\}$. The *direct product* of $A_1, A_2, \ldots, A_n$, written

$$\prod_{i=1}^{n} A_i$$

(read "product i equals one to n of A_i") is the set consisting of all sequences $(a_1, a_2, \ldots, a_n)$ such that $a_1 \in A_1, a_2 \in A_2, \ldots, a_n \in A_n$. In particular,

$$\prod_{i=1}^{2} A_i = A_1 \times A_2.$$

For this reason we shall often write

$$A_1 \times A_2 \times \ldots \times A_n$$

for $\prod_{i=1}^{n} A_i$.

The concept of direct product may be extended to an infinite sequence $A_1, A_2, \ldots, A_n, \ldots$ of sets, indexed by the positive integers. The *direct product* of $A_1, A_2, \ldots, A_n, \ldots$, written

$$\prod_{i=1}^{\infty} A_i$$

or

$$A_1 \times A_2 \times \ldots \times A_n \times \ldots$$

is the set whose elements are all infinite sequences $(a_1, a_2, \ldots, a_n, \ldots)$ such that $a_i \in A_i$ for each positive integer i.

The set of points of Euclidean n-space yields an example of a direct product of sets. If for $i = 1, 2, \ldots, n$ we have $A_i = R$, where R is the set of real numbers, then

$$R^n = \prod_{i=1}^{n} A_i$$

is the set of points of a Euclidean n-space. An element $x \in R^n$ is a sequence $x = (x_1, x_2, \ldots, x_n)$ of real numbers. In general, if the sets $A_1, A_2, \ldots, A_n$ are all equal to the same set A, we write

$$A^n = \prod_{i=1}^{n} A_i$$

and call an element $a = (a_1, a_2, \ldots, a_n) \in A^n$ an n-tuple.

EXERCISES

1. Let $X \subset A$, $Y \subset B$. Prove that
$$C(X \times Y) = A \times C(Y) \cup C(X) \times B.$$
2. Prove that if A has precisely n distinct elements and B has precisely m distinct elements, where m and n are positive integers, then $A \times B$ has precisely mn distinct elements.
3. Let A and B be sets, both of which have at least two distinct members. Prove that there is a subset $W \subset A \times B$ that is not the Cartesian product of a subset of A with a subset of B. [Thus, not every subset of a Cartesian product is the Cartesian product of a pair of subsets.]

6 FUNCTIONS

The most familiar example of a function in mathematics is a correspondence that associates with each real number x a real number $f(x)$. The purpose of marking an examination may be described as the construction of a marking function that makes correspond to each student taking the examination some integer between zero and one hundred. Integration of a continuous function defined on some closed interval $[a, b]$ is another example of a function, namely the correspondence that associates with each object f in this given set of objects the real number

$$\int_a^b f(x)\,dx.$$

The concept of function or correspondence need not be restricted to the realm of numerical quantities. The correspondence that associates with each undergraduate in college one of the four adjectives *freshman, sophomore, junior,* or *senior* is also an example of a function using correspondence as an undefined concept.

DEFINITION Let A and B be sets. A correspondence that associates with each element $x \in A$ a unique element $f(x) \in B$ is called a *function* from A to B, which we shall write

$$f: A \to B,$$

or

$$A \xrightarrow{f} B.$$

DEFINITION Let $f: A \to B$. The subset $\Gamma_f \subset A \times B$, which consists of all ordered pairs of the form $(a, f(a))$ is called the *graph* of $f: A \to B$.

The graph Γ_f of a function $f: X \to Y$ is the subset of $X \times Y$ consisting of precisely those points (x, y) for which the statement $f(x) = y$ is true. This set is sometimes written

$$\{(x, y) \mid (x, y) \in X \times Y \quad \text{and} \quad y = f(x)\}.$$

This notation, called the set builder notation, is of the general form $\{z \mid P(z)\}$, where $P(z)$ is some statement which may or may not be true of z. The resulting set is the set of all z, in an appropriate universe, for which $P(z)$ is true.

Let A and B be sets. Given a subset Γ of $A \times B$ there is a function $f: A \to B$ such that Γ is the graph of $f: A \to B$ if, for each $x \in A$, there is one and only one element of the form $(x, y) \in \Gamma$.

(Thus the equivalent definition of a function as a subset $\Gamma \subset A \times B$ with the aforesaid property is frequently employed, in which case for each $x \in A$, the function Γ makes correspond to x the element $y \in B$ such that $(x, y) \in \Gamma$.)

DEFINITION Let $f:A \to B$ be given. For each subset X of A, the subset of B whose elements are the points $f(x)$ such that $x \in X$ is denoted by $f(X)$. $f(X)$ is called the *image* of X. For each subset Y of B, the subset of A whose elements are the points $x \in A$ such that $f(x) \in Y$ is denoted by $f^{-1}(Y)$. $f^{-1}(Y)$ is called the *inverse image* of Y, *counter image* of Y, or f *inverse* of Y.

DEFINITION Let $f:A \to B$ be given. A is called the *domain* of f. B is called the *range* of f.

EXAMPLE Let $f:R \to R$, R the set of real numbers, be the function such that for each $x \in R$, $f(x) = x^2 - x - 2$. If X is the closed interval $[1, 2]$, then $f(X) = [-2, 0]$. If Z is the open interval $(-1, 1)$, then $f(Z) = (-9/4, 0) \cup \{-9/4\}$. $f^{-1}([-2, 0]) = [1, 2] \cup [-1, 0]$. $f^{-1}(\{0\}) = \{2, -1\}$ is the set of roots of the polynomial $x^2 - x - 2$. $f^{-1}([-5, -4]) = \emptyset$.

A function $f:A \to B$ is also called a *mapping* or *transformation* of A into B. We may think of such a function as carrying each point $x \in A$ into its corresponding point $f(x) \in B$.

DEFINITION A function $f:A \to B$ is called *one-one* if whenever $f(a) = f(a')$ for $a, a' \in A$, then $a = a'$.

Thus, $f:A \to B$ is one-one if for each $b \in f(A)$ there is only one $a \in A$ such that $f(a) = b$.

DEFINITION A function $f:A \to B$ is called *onto* if $B = f(A)$.

A one-one function is sometimes called *injective* and an onto function is sometimes called *surjective*. A function which is both one-one and onto is sometimes called *bijective*.

DEFINITION A function $f: A \to B$ is called a *constant* function if there is a point $b \in B$ such that $f(x) = b$ for each $x \in A$.

DEFINITION A function $f: A \to A$ is called the *identity* function (on A) if $f(x) = x$ for each $x \in A$.

EXERCISES

1. Let $f: A \to B$ be given. Prove the following:
   (a) For each subset $X \subset A$, $X \subset f^{-1}(f(X))$.
   (b) For each subset $Y \subset B$, $Y \supset f(f^{-1}(Y))$.
   (c) If $f: A \to B$ is one-one, then for each subset $X \subset A$,
   $$f^{-1}(f(X)) = X.$$
   (d) If $f: A \to B$ is onto, then for each subset $Y \subset B$,
   $$f(f^{-1}(Y)) = Y.$$
2. Let $A = \{a_1, a_2\}$ and $B = \{b_1, b_2\}$ be two sets, each having precisely two distinct elements. Let $f: A \to B$ be the constant function such that $f(a) = b_1$ for each $a \in A$.
   (a) Prove that $f^{-1}(f(\{a_1\})) \neq \{a_1\}$. [Thus it is usually the case that $f^{-1}(f(X))$ and X are not equal.]
   (b) Prove that $f(f^{-1}(B)) \neq B$. [Thus it is usually the case that $f(f^{-1}(B))$ and B are not equal.]
   (c) Prove that $f(\{a_1\} \cap \{a_2\}) \neq f(\{a_1\}) \cap f(\{a_2\})$. [Thus it is usually the case that $f(X \cap X')$ and $f(X) \cap f(X')$ are not equal.]
3. Let $f: A \to B$ be given and let $\{X_\alpha\}_{\alpha \in I}$ be an indexed family of subsets of A. Prove:
   (a) $f(\bigcup_{\alpha \in I} X_\alpha) = \bigcup_{\alpha \in I} f(X_\alpha)$.
   (b) $f(\bigcap_{\alpha \in I} X_\alpha) \subset \bigcap_{\alpha \in I} f(X_\alpha)$.
   (c) If $f: A \to B$ is one-one, then $f(\bigcap_{\alpha \in I} X_\alpha) = \bigcap_{\alpha \in I} f(X_\alpha)$.
4. Let $f: A \to B$ be given and let $\{Y_\alpha\}_{\alpha \in I}$ be an indexed family of subsets of B. Prove:
   (a) $f^{-1}(\bigcup_{\alpha \in I} Y_\alpha) = \bigcup_{\alpha \in I} f^{-1}(Y_\alpha)$.
   (b) $f^{-1}(\bigcap_{\alpha \in I} Y_\alpha) = \bigcap_{\alpha \in I} f^{-1}(Y_\alpha)$.
   (c) If X is a subset of B then $f^{-1}(C(X)) = C(f^{-1}(X))$.
   (d) If X is a subset of A, and Y is a subset of B, then $f(X \cap f^{-1}(Y)) = f(X) \cap Y$.

5. Let A and B be sets. The correspondence that associates with each element $(a, b) \in A \times B$ the element $p_1(a, b) = a$ is a function called the *first projection*. The correspondence that associates with each element $(a, b) \in A \times B$ the element $p_2(a, b) = b$ is a function called the *second projection*. Prove that if $B \neq \emptyset$, then $p_1: A \times B \to A$ is onto and if $A \neq \emptyset$, then $p_2: A \times B \to B$ is onto. Under what circumstances is p_1 or p_2 one-one? What is $p_1^{-1}(\{a\})$ for an element $a \in A$?

6. Let A and B be sets, with $B \neq \emptyset$. For each $b \in B$ the correspondence that associates with each element $a \in A$ the element $j_b(a) = (a, b) \in A \times B$ is a function. Prove that for each $b \in B$, $j_b: A \to A \times B$ is one-one. What is $j_b^{-1}(W)$ for a subset $W \subset A \times B$?

7. Let A be a set and $E \subset A$. The function $\chi_E: A \to \{0, 1\}$ defined by $\chi_E(x) = 1$ if $x \in E$ and $\chi_E(x) = 0$ if $x \notin E$ is called the *characteristic* function of E. Let E and F be subsets of A, show:
   (a) $\chi_{E \cap F} = \chi_E \cdot \chi_F$, where $\chi_E \cdot \chi_F(x) = \chi_E(x)\chi_F(x)$;
   (b) $\chi_{E \cup F} = \chi_E + \chi_F - \chi_{E \cap F}$ and find a similar expression for $\chi_{E \cup F \cup G}$.

8. Let A be a set to which there belong precisely n distinct objects. Prove that there are precisely 2^n distinct objects that belong to 2^A.

7 RELATIONS

A function may be viewed as a special case of what is called a relation. We are accustomed to thinking of one object being in a given relation to another; for example, Jeanne is the sister of Sam or silk purses are more expensive than sows' ears. To say that the number 2 is less than the number 3, or $2 < 3$, is thus to say that $(2, 3)$ is one of the number pairs (x, y) for which the relation "less than" is true.

DEFINITION A *relation* R from the elements of a set A to the elements of a set B is a subset of $A \times B$. A *relation* R on a set E is a subset of $E \times E$.

If $(x, y) \in R \subset A \times B$, one frequently writes aRb. We wish to distinguish certain properties that a relation on a set E may or may not have.

DEFINITION A relation R on a set E is called *reflexive* if aRa is true for all $a \in E$. It is called *symmetric* if, whenever aRb, also bRa. It is called *transitive* if, whenever aRb and bRc, then aRc.

Let $<$ be the pairs of real numbers (x, y) such that $x < y$. Then $<$ is a transitive relation on the set E of real numbers, but $<$ is not reflexive and not symmetric. Let R be the pairs of real numbers (x, y) such that $|x - y| < 1$. Then R is reflexive and symmetric, but not transitive. Let Λ be the pairs of real numbers (x, y) such that $x - y$ is an integer. Then Λ is reflexive, symmetric, and transitive.

DEFINITION A relation R on a set E which is reflexive, symmetric, and transitive is called an *equivalence* relation.

DEFINITION Let R be an equivalence relation on a set E. For each $a \in E$, the *equivalence class* of a, denoted by $\pi(a)$, is the subset of E consisting of all x such that aRx.

Two equivalence classes are either disjoint or identical.

LEMMA Let R be an equivalence relation on a set E and let $\pi(a) \cap \pi(b) \neq \varnothing$ for $a, b \in E$. Then $\pi(a) = \pi(b)$.

Proof. Let $c \in \pi(a)$, $c \in \pi(b)$. Then aRc and bRc. Suppose $x \in \pi(a)$ so that aRx. cRa by symmetry, so cRx by transitivity. Another application of transitivity yields bRx, so $x \in \pi(b)$. Thus $\pi(a) \subset \pi(b)$. Similarly $\pi(b) \subset \pi(a)$.

By the reflexive property $a \in \pi(a)$ is always true, so the equivalence classes are non-empty and disjoint. Let E/R be the set of equivalence classes, then $\pi : E \to E/R$ is an onto function. E/R is sometimes called the *quotient* of E by the relation R, and π is called the *projection*.

EXERCISES

1. Let P be a subset of the real numbers R such that (i) $1 \in P$, (ii) if $a, b \in P$ then $a + b \in P$, and (iii) for each $x \in R$, one and only one of the three statements, $x \in P$, $x = 0$, or $-x \in P$ is true. Define $Q = \{(a, b) \mid (a, b) \in R \times R \text{ and } a - b \in P\}$. Prove that Q is a transitive relation.
2. Let $f:X \to Y$ be a function from a set X onto a set Y. Let R be the subset of $X \times X$ consisting of those pairs (x, x') such that $f(x) = f(x')$. Prove that R is an equivalence relation. Let $\pi:X \to X/R$ be the projection. Verify that, if $\alpha \in X/R$ is an equivalence class, to define $F(\alpha) = f(a)$, whenever $\alpha = \pi(a)$, establishes a well-defined function $F:X/R \to Y$ which is one-one and onto.
3. Let $f:X \to X$ be a one-one function of a set X into itself. Define a sequence of functions $f^0, f^1, f^2, \cdots, f^n, \cdots :X \to X$ by letting f^0 be the identity, $f^1 = f$, and inductively $f^n(x) = f(f^{n-1}(x))$. Prove that each of these functions is one-one. Let R be the subset of $X \times X$ consisting of those pairs (a, b) such that $b = f^k(a)$ for some integer k or $a = f^j(b)$ for some integer j. Prove that R is an equivalence relation.
4. Let X be the set of functions from the real numbers into the real numbers possessing continuous derivatives. Let R be the subset of $X \times X$ consisting of those pairs (f, g) such that $Df = Dg$ where D maps a function into its derivative. Prove that R is an equivalence relation and describe an equivalence set $\pi(f)$.
5. Let E be the set of all functions from a set X into a set Y. Let $b \in X$ and let R be the subset of $E \times E$ consisting of those pairs (f, g) such that $f(b) = g(b)$. Prove that R is an equivalence relation. Define a one-one onto function $e_b:E/R \to Y$.

8 COMPOSITION OF FUNCTIONS AND DIAGRAMS

Definition Let $f:A \to B$ and $g:B \to C$ be given. The *composition* of $f:A \to B$ and $g:B \to C$ is the correspondence that associates with each element $a \in A$, the element $g(f(a)) \in C$. This function is written $gf:A \to C$, or $A \xrightarrow{gf} C$.

A function $h:A \to C$ is, therefore, the composition of $f:A \to B$ and $g:B \to C$ (often abbreviated by writing $h = gf$) if for each $a \in A$, $h(a) = g(f(a))$. In a pictorial representation of these functions, we have $h = gf$ when these functions behave in the manner indicated in Figure 2.

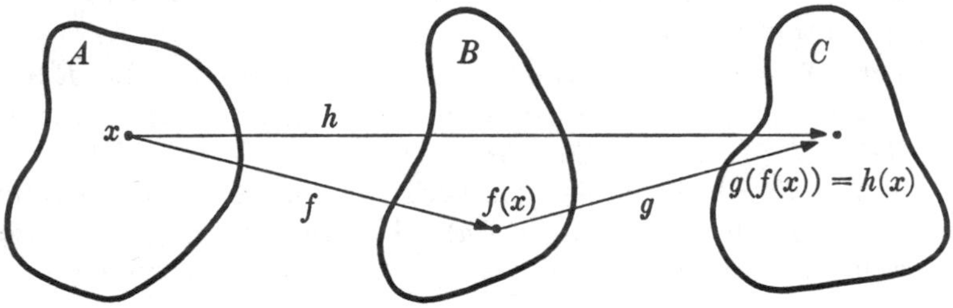

Figure 2

The concept of the composition of functions can be extended to the composition of a finite number of functions.

DEFINITION Let $f_1:A_1 \to A_2$, $f_2:A_2 \to A_3$, . . . , $f_n:A_n \to A_{n+1}$ be given. The *composition* of $f_1:A_1 \to A_2$, $f_2:A_2 \to A_3$, . . . , and $f_n:A_n \to A_{n+1}$ is the correspondence that associates with each element $x \in A_1$ the element $f_n(\ldots f_2(f_1(x))\ldots) \in A_{n+1}$. This function is written

$$f_n \ldots f_2 f_1 : A_1 \to A_{n+1},$$

or

$$A_1 \xrightarrow{f_n \ldots f_2 f_1} A_{n+1}.$$

Let three functions $f:A \to B$, $g:B \to C$, and $h:C \to D$ be given. We may form $hgf:A \to D$. We may also form $gf:A \to C$ and compose this function with $h:C \to D$ to obtain $h(gf):A \to D$. Similarly, we may form $(hg)f:A \to D$. We thus have three functions hgf, $h(gf)$, $(hg)f:A \to D$. But

$$(hgf)(x) = h(g(f(x)));$$

$$(h(gf))(x) = h((gf)(x)) = h(g(f(x)));$$

$$((hg)f)(x) = (hg)(f(x)) = h(g(f(x))).$$

Thus, these three functions are the same. This observation provides a basis for the justification of the removal or replacement of parentheses in expressions such as $(f_4f_3)(f_2f_1)$, etc.

Suppose we are given three functions $f:A \to B$, $g:B \to C$, and $k:A \to C$. The existence of these three functions may be indicated, as in Figure 3, by what we shall call a *diagram*. The letters A, B, C stand for the various sets, and an arrow leading

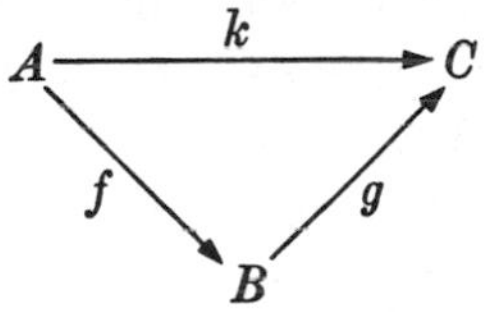

Figure 3

from one set to another indicates a function from the first set to the second, namely, the function that carries each element x of the first set into the element $t(x)$ of the second set, where t stands for the symbol closest to the middle of the arrow. The fact that we may form the composition of two functions (such as $gf:A \to C$ in the above diagram) is represented by a path in the direction of the arrows that goes from one set to a second and from the second set to a third. (In the above diagram we say, "We may go from A to B via f and from B to C via g.")

We shall desire to diagram more complex situations than the one indicated in Figure 3. Let us say that by a *diagram* we shall mean a figure consisting of several symbols denoting sets and arrows leading from one symbol to another, each arrow leading from a set X to a set Y having an associated symbol t, the arrow and its symbol representing a given function $t:X \to Y$. For example, diagram (8.1) indicates the existence of given functions $f:A \to B$, $g:A \to C$, $k:B \to D$, $h:C \to D$. This diagram shows us

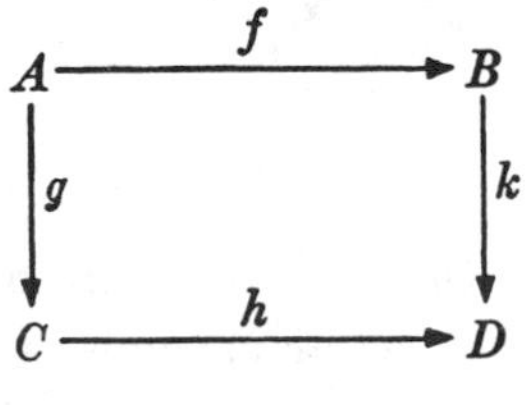

(8.1)

that by composing functions we may obtain two functions from A to D, namely, kf, $hg: A \to D$. In any diagram, a path from X to Y consisting of a sequence of arrows leading from X to Y indicates the existence of a function from X to Y obtained by composing the functions represented by these arrows in the order of their occurrence, starting at X and terminating at Y.

In diagram (8.1) it may or may not be true that $kf = hg$. In the event that $kf = hg$ we will say that diagram (8.1) is *commutative*. In general, a diagram is said to be *commutative* if for each X and Y in the diagram that represent sets, and for any two paths in the diagram beginning at X and ending at Y, the two functions from X to Y so represented are equal. For example, the statement that diagram (8.2) is commutative means that $f = jh$, $k = gj$, and $kh = gjh = gf$ (note that the first two equalities imply the third).

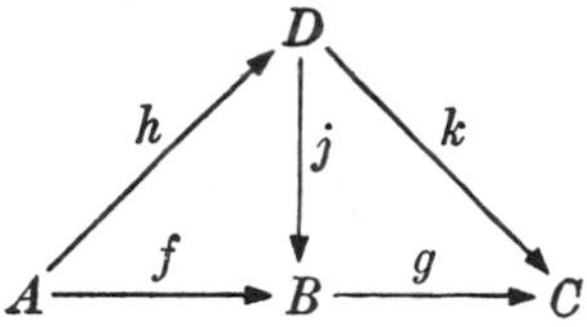

(8.2)

A given set A may occur more than once in a diagram. For example, let A be the set of positive real numbers and R the set of real numbers. Let $f: A \to R$ be defined by the correspondence $f(x) = \log_e x$, $x \in A$, and let $g: R \to A$ be defined by the correspondence $g(x) = e^x$, $x \in R$. Let $i: A \to A$ be the identity function. Then the diagram (8.3) is commutative, for $(gf)(x) = e^{\log_e x} = x = i(x)$ for $x \in A$.

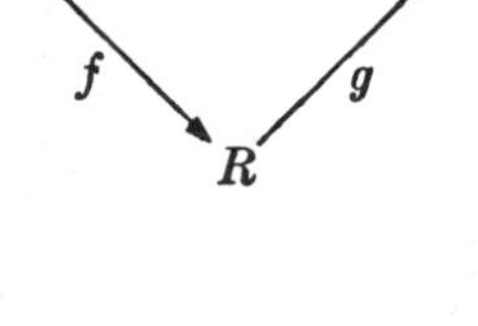

(8.3)

EXERCISES

1. Using the functions defined by the correspondences $g(x) = x^2$ and $h(x) = \sqrt{x}$, $x \geqq 0$, construct an example of a commutative diagram

similar to diagram (8.3).

2. Let $f:R \times R \to R$ be the function defined by the correspondence $f(x, y) = x^2 + y^2$ and let $g:R \times R \to R$ be the function defined by the correspondence $g(x, y) = x + y$. Let $h:R \to R$ be the function defined by the correspondence $h(x) = x^2$. Is the diagram

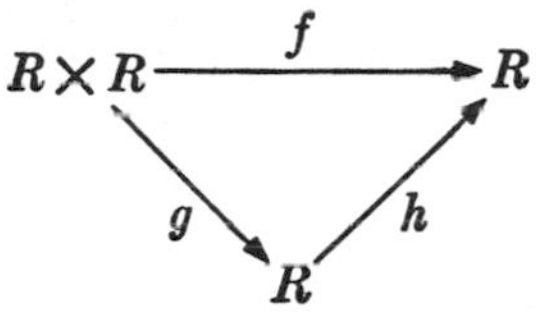

commutative?

3. Let $i:A \to A$ be the identity function. Let the diagram

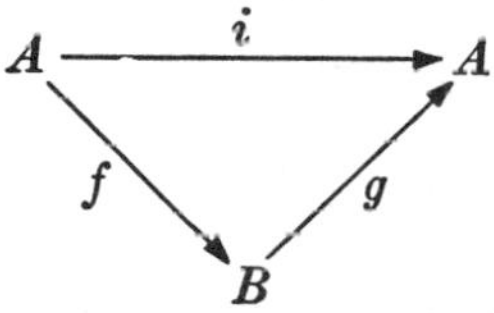

be commutative. Prove that $g:B \to A$ is onto and that $f:A \to B$ is one-one.

4. Let $f:A \to B$, $g:B \to C$. Prove that for $Z \subset C$, $(gf)^{-1}(Z) = f^{-1}(g^{-1}(Z))$.

9 INVERSE FUNCTIONS, EXTENSIONS, AND RESTRICTIONS

DEFINITION Let $f:A \to B$ and $g:B \to A$ be given. The function $f:A \to B$ is called the *inverse* of $g:B \to A$ and the function $g:B \to A$ is called the *inverse* of $f:A \to B$ if $g(f(a)) = a$ for each $a \in A$ and $f(g(b)) = b$ for each $b \in B$.

In this event we shall also say that $f:A \to B$ and $g:B \to A$ are *inverse functions* and that each of them is *invertible.*

Let $i_A: A \rightarrow A$ and $i_B: B \rightarrow B$ be identity functions. The statement that $f: A \rightarrow B$ and $g: B \rightarrow A$ are inverse functions is equivalent to the statement that the two diagrams

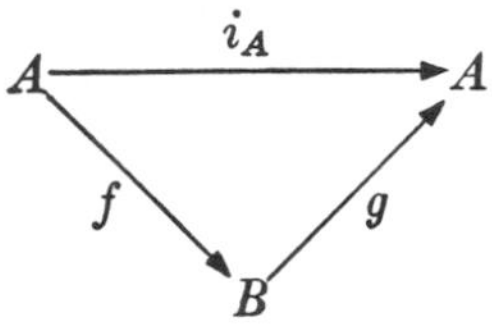

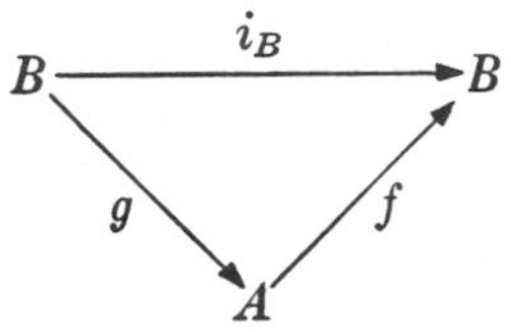

are commutative.

Theorem Let $f: A \rightarrow B$ and $g: B \rightarrow A$ be inverse functions, then both functions are one-one and onto.

Proof. Suppose $f(x) = f(y)$, x, $y \in A$. Then $x = g(f(x)) = g(f(y)) = y$ and therefore f is one-one. To show that f is onto, let $b \in B$. We have $g(b) \in A$ and $f(g(b)) = b$, therefore if we set $a = g(b)$, $b = f(a)$ and f is onto. The roles of the two functions may be interchanged, since the definition of inverse functions imposes conditions symmetrical with regard to the two functions. Therefore, $g: B \rightarrow A$ is also one-one and onto.

We have shown that, given a function $h: X \rightarrow Y$, a necessary condition that this function be invertible is that the function be one-one and onto. This condition is also sufficient.

Theorem Let $f: A \rightarrow B$ be one-one and onto. Then there exists a function $g: B \rightarrow A$ such that these two functions are inverse functions.

Proof. We shall first define $g: B \rightarrow A$. Given $b \in B$, we may write $b = f(a)$ for some $a \in A$ since f is onto. Furthermore, f is one-one; hence there is only one element $a \in A$ such that $f(a) = b$. We define $g(b) = a$. The correspondence that associates with each $b \in B$ the element $g(b) \in A$, as defined above, is a function $g: B \rightarrow A$. $f(g(b)) = b$ for each $b \in B$ by the definition of $g: B \rightarrow A$. Given $a \in A$, let $a' = g(f(a))$. Then $f(a') = f(g(f(a))) = f(a)$ by the remark just made. Since $f:A \to B$ is one-one, $a = a' = g(f(a))$. Thus, $f:A \to B$ and $g:B \to A$ are inverse functions.

The last two theorems may be combined in the statement: given $f:A \to B$, a necessary and sufficient condition that there be a function $g:B \to A$ such that these two functions are inverse functions is that $f:A \to B$ be one-one and onto. Furthermore, in this event, the function $g:B \to A$ is uniquely determined.

Theorem Let $f:A \to B$, $g:B \to A$ be inverse functions and let $f:A \to B$ and $g':B \to A$ be inverse functions. Then $g:B \to A$ and $g':B \to A$ are equal.

Proof. We must prove that $g(b) = g'(b)$ for each $b \in B$. But $b = f(g(b))$ and therefore $g'(b) = g'(f(g(b))) = g(b)$, since $g'(f(a)) = a$ for each $a \in A$.

The proof of this last theorem may also be viewed as a direct consequence of the commutativity of the diagram

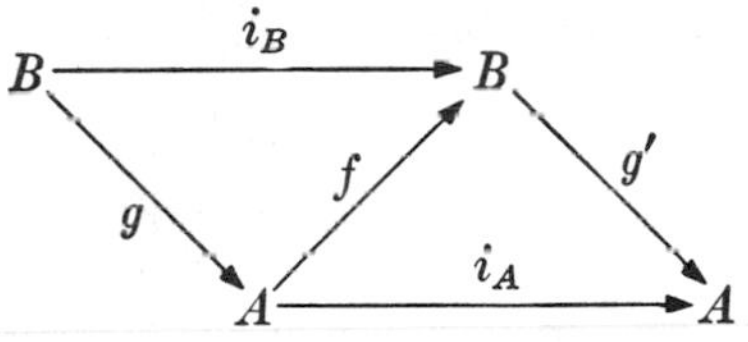

which yields $g'(b) = g'(i_B(b)) = g'(f(g(b))) = i_A(g(b)) = g(b)$.

Definition Let $A \subset X$. Let $f:A \to Y$ and $F:X \to Y$. If for each $x \in A$, $f(x) = F(x)$, we say that *F is an extension of f to X* or that *f is a restriction of F to A*. In this event we shall write $f = F \mid A$.

Example Let A be the open interval $(0, \pi/2)$. For each $\theta \in A$, let Δ_θ be a right triangle one of whose acute angles is θ radians, and let $f(\theta)$ be the ratio of the length of the side of this triangle opposite the angle of magnitude θ to the length of the hypotenuse of Δ_θ, (more familiarly,

$$f(\theta) = \frac{\text{opposite}}{\text{hypotenuse}}\Big).$$

Thus $f:A \to R$. For each $\theta \in R$, let $(a, b)_\theta$ be the point of the plane R^2 whose distance from the origin is 1 and such that the rotation about the origin of the line segment whose end points are the origin and $(1, 0)$ to the position of the line segment whose end points are the origin and $(a, b)_\theta$ represents an angle of magnitude θ radians. Define $F(\theta) = b$. Then $F:R \to R$. F is an extension of f to R as is easily seen if one recognizes $f:A \to R$ as the sine function defined for acute angles by means of right triangles and $F:R \to R$ as the sine function defined for angles of arbitrary magnitude by means of the unit circle.

Definition Let $A \subset X$. The function $i:A \to X$, which is defined by the correspondence $i(x) = x$ for each $x \in A$ is called an *inclusion* mapping or function.

Let $A \subset X, f:A \to Y$ and $F:X \to Y$. Then F is an extension of f if and only if the diagram

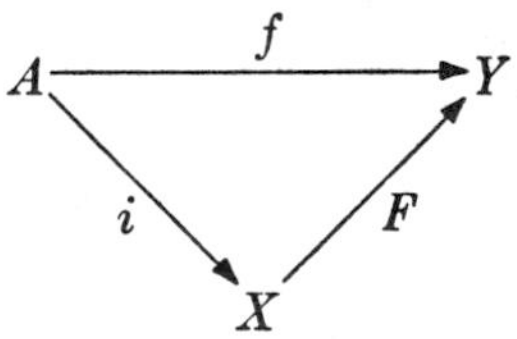

is commutative, where $i:A \to X$ is an inclusion mapping.

Given $F:X \to Y$, there are as many restrictions of $F:X \to Y$ as there are subsets of X. Given a subset $A \subset X$, we may obtain the restriction of F to A by forming the composition of the inclusion mapping $i:A \to X$ and $F:X \to Y$. Thus, we may write $F \mid A = Fi$.

EXERCISES

1. Let A be the set of all functions $f:[a, b] \to R$ that are continuous on $[a, b]$. Let B be the subset of A consisting of all functions possessing a continuous derivative on $[a, b]$. Let C be the subset of B consisting of all functions whose value at a is 0. Let $d:B \to A$ be the correspondence that associates with each function in B its derivative. Is the function $d:B \to A$ invertible?

   To each $f \in A$, let $h(f)$ be the function defined by

   $$(h(f))(x) = \int_a^x f(t)\,dt,$$

   for $x \in [a, b]$. Verify that $h:A \to C$. Find the function $g:C \to A$ such that these two functions are inverse functions.
2. Let R be the real numbers and ∞ an object not in R. Define a set $R^* = R \cup \{\infty\}$. Let a, b, c, d be real numbers. Let $f:R^* \to R^*$ be a function defined by $f(x) = (ax + b)/(cx + d)$ when $x \neq -d/c, \infty$, while $f(-d/c) = \infty$ and $f(\infty) = a/c$. [In the event that $c = 0$, f is linear and $f(x) = (ax + b)/d$ when $x \neq \infty$ and $f(\infty) = \infty$.] Prove that f has an inverse provided $ad - bc \neq 0$.
3. Let $A \subset B \subset X$. Let $f:A \to Y$, $g:B \to Y$, and $F:X \to Y$. Prove that if g is an extension of f to B and F is an extension of g to X, then F is an extension of f to X.
4. Let m, n be positive integers. Let X be a set with m distinct elements and Y a set with n distinct elements. How many distinct functions are there from X to Y? Let A be a subset of X with r distinct elements, $0 \leq r < m$ and $f:A \to Y$. How many distinct extensions of f to X are there?

10 ARBITRARY PRODUCTS

Let $X_1, \ldots, X_n$ be sets. We have defined a point

$$x = (x_1, \ldots, x_n) \in \prod_{i=1}^{n} X_i$$

as an ordered sequence such that $x_i \in X_i$. Given such a point, by setting $x(i) = x_i$ we obtain a function x which associates to each integer i, $1 \leqq i \leqq n$, the element $x(i) \in X_i$. Conversely, given a function x which associates to each integer i, $1 \leqq i \leqq n$, an element $x(i) \in X_i$, we obtain the point

$$(x(1), \ldots, x(n)) \in \prod_{i=1}^{n} X_i$$

It is easily seen that this correspondence between points of $\prod_{i=1}^{n} X_i$ and functions of the above type is one-one and onto, so that a point of $\prod_{i=1}^{n} X_i$ may also be defined as a function x which associates to each integer i, $1 \leqq i \leqq n$, a point $x(i) \in X_i$. The advantage of this second approach is that it allows us to define the product of an arbitrary family of sets.

DEFINITION Let $\{X_\alpha\}_{\alpha \in I}$ be an indexed family of sets. The product of the sets $\{X_\alpha\}_{\alpha \in I}$, written $\Pi_{\alpha \in I} X_\alpha$, consists of all functions x with domain the indexing set I having the property that for each $\alpha \in I$, $x(\alpha) \in X_\alpha$.

Given a point $x \in \Pi_{\alpha \in I} X_\alpha$, one may refer to $x(\alpha)$ as the αth coordinate of x. However, unless the indexing set has been ordered in some fashion (as is the case with finite products in our earlier discussion), there is no first coordinate, second coordinate, and so on.

DEFINITION Let $x \in \Pi_{\alpha \in I} X_\alpha$. The function $p_\alpha : \Pi_{\alpha \in I} X_\alpha \to X_\alpha$ defined by $p_\alpha(x) = x(\alpha)$ is called the αth *projection.*

Clearly two points $x, x' \in \Pi_{\alpha \in I} X_\alpha$ are identical if and only if, for each $\alpha \in I$, $p_\alpha(x) = p_\alpha(x')$, that is, $x(\alpha) = x'(\alpha)$.

In dealing with product spaces use is frequently made of a principle, called the *axiom of choice,* whereby we assume that if for each $\alpha \in I$ we can choose a point $x_\alpha \in X_\alpha$, then we may construct a point or function $x \in \Pi_{\alpha \in I} X_\alpha$ by setting $x(\alpha) = x_\alpha$. This is equivalent to the statement that the product of non-empty sets is non-empty. Using the axiom of choice we may prove

PROPOSITION If for each $\alpha \in I$, X_α is non-empty, then each of the projection maps $p_\alpha : \Pi_{\alpha \in I} X_\alpha \to X_\alpha$ is onto.

Proof. Let $x_\alpha \in X_\alpha$ be given. Set $x(\alpha) = x_\alpha$. Suppose $\beta \in I$, $\beta \neq \alpha$. Since X_β is non-empty we may choose a point $x(\beta) \in X_\beta$. Then $x \in \Pi_{\alpha \in I} X_\alpha$ and $p_\alpha(x) = x(\alpha) = x_\alpha$, hence p_α is onto.

In the above proof we have obtained a point $x \in p_\alpha^{-1}(x_\alpha)$, that is, a point whose α^{th} coordinate is x_α and whose other coordinates are unrestricted. If $B \subset X_\alpha$ then to say that $x \in p_\alpha^{-1}(B)$ is to restrict the αth coordinate of x to lie in B and leave all other coordinates unrestricted.

EXERCISES

1. Let A be a set. For each $\alpha \in I$, let $X_\alpha = A$. Verify that $\Pi_{\alpha \in I} X_\alpha$ is the set of all functions from the set I to the set A. This set of functions is denoted by A^I. Suppose $A = \{0, 1\}$. If I is finite how many elements are there in A^I? Verify that A^I in this case is the set of all characteristic functions defined on I.
2. Let $\{X_\alpha\}_{\alpha \in I}$, $\{Y_\alpha\}_{\alpha \in I}$ be two indexed families of sets with the same indexing set I. For each $\alpha \in I$ let $f_\alpha : X_\alpha \to Y_\alpha$. Prove that there is a unique function $f : \Pi_{\alpha \in I} X_\alpha \to \Pi_{\alpha \in I} Y_\alpha$ such that $p_\alpha f = f_\alpha p_\alpha$ for each $\alpha \in I$, where p_α is the appropriate projection map. This function f is denoted by $\Pi_{\alpha \in I} f_\alpha$. Given a third indexed family of sets $\{Z_\alpha\}_{\alpha \in I}$ and functions $g_\alpha : Y_\alpha \to Z_\alpha$ for each $\alpha \in I$, show that $\Pi_{\alpha \in I} g_\alpha \Pi_{\alpha \in I} f_\alpha = \Pi_{\alpha \in I} g_\alpha f_\alpha$. Suppose that each f_α has an inverse k_α. Prove that $\Pi_{\alpha \in I} f_\alpha$ has the inverse $\Pi_{\alpha \in I} k_\alpha$.
3. Let $\{X_\alpha\}_{\alpha \in I}$ be an indexed family of sets and let $I = I_1 \cup I_2$, where $I_1 \cap I_2 = \emptyset$. Show that there is a one-one mapping of $(\Pi_{\alpha \in I_1} X_\alpha) \times (\Pi_{\alpha \in I_2} X_\alpha)$ onto $\Pi_{\alpha \in I} X_\alpha$. More generally, let $\{I_\gamma\}_{\gamma \in J}$ be a partition of I, that is $I = \bigcup_{\gamma \in J} I_\gamma$, $I_{\gamma_1} \cap I_{\gamma_2} = \emptyset$ for $\gamma_1 \neq \gamma_2$, each $I_\gamma \neq \emptyset$. Show that there is a one-one mapping of $\Pi_{\gamma \in J} (\Pi_{\alpha \in I_\gamma} X_\alpha)$ onto $\Pi_{\alpha \in I} X_\alpha$.
4. Let N be the set of positive integers. In the notation of Problem 1, an infinite sequence $x_1, x_2, \ldots$ of points of a set X may be viewed as an element $x \in X^N$. If $j : N \to N$ is a function such that $j(i) < j(i + 1)$ for $i \in N$, then the infinite sequence xj is a sub-

sequence of the sequence x. Prove that a subsequence of xj is a subsequence of x.

For further reading, the books by Halmos, *Naive Set Theory*, and Kaplansky, *Set Theory and Metric Spaces* are both excellent sources.

CHAPTER 2

Metric Spaces

1 INTRODUCTION

A metric space is a set of points and a prescribed quantitative measure of the degree of closeness of pairs of points in this space. The real number system and the coordinate plane of analytic geometry are familiar examples of metric spaces. Starting from the vague characterization of a continuous function as one that transforms nearby points into points that are themselves nearby, we can, in a metric space, formulate a precise definition of continuity. Although this definition may be stated in the so-called "ε, δ" terminology, there are other, equivalent formulations available in a metric space. These include characterizations of continuity in terms of the behavior of a function with respect to certain subsets called neighborhoods of a point, or with respect to certain subsets called open sets.

2 METRIC SPACES

Given two real numbers a and b, there is determined a non-negative real number, $|a - b|$, called the distance between a and b. Since to each ordered pair (a, b) of real numbers there is associated the real number $|a - b|$, we may write this correspondence in functional notation by setting

$$d(a, b) = |a - b|.$$

Thus we have a function $d: R \times R \to R$, where R is the set of real numbers. This function has four important properties, which the reader should verify:

1. $d(x, y) \geqq 0$;
2. $d(x, y) = 0$ if and only if $x = y$;
3. $d(x, y) = d(y, x)$;
4. $d(x, z) \leqq d(x, y) + d(y, z)$;

for $x, y, z \in R$. For the purposes of discussing "continuity" of functions, these four properties of "distance" are sufficient. This fact suggests the possibility of examining "continuity" in a more general setting; namely, in terms of any set of points for which there is defined a "distance function" such as the function $d: R \times R \to R$ above.

DEFINITION 2.1 A pair of objects (X, d) consisting of a non-empty set X and a function $d: X \times X \to R$, where R is the set of real numbers, is called a *metric space* provided that:

1. $d(x, y) \geqq 0, \quad x, y \in X$;
2. $d(x, y) = 0$ if and only if $x = y, \quad x, y \in X$;
3. $d(x, y) = d(y, x), \quad x, y \in X$;
4. $d(x, z) \leqq d(x, y) + d(y, z), \quad x, y, z \in X$.

The function d is called a *distance function* or *metric* on X and the set X is called the *underlying set.*

[A more precise notation for a metric space would be $(X, d:X \times X \to R)$ and for a distance function $d:X \times X \to R$. We shall, however, frequently delete the sets and arrow in the symbol for a function, when, in a given context, it is clear which sets are involved.]

We may think of the distance function d as providing a quantitative measure of the degree of closeness of two points. In particular, the inequality $d(x, z) \leqq d(x, y) + d(y, z)$ may be thought of as asserting the transitivity of closeness; that is, if x is close to y and y is close to z, then x is close to z.

Let $a, b \in R$, where R is the set of real numbers. The verification that the function $d(a, b) = |a - b|$ satisfies the four properties enumerated in Definition 2.1 establishes:

THEOREM 2.2 (R, d) is a metric space, where d is the function defined by the correspondence $d(a, b) = |a - b|$, for $a, b \in R$.

Given a finite collection $(X_1, d_1), (X_2, d_2), \ldots, (X_n, d_n)$ of metric spaces, there is a standard procedure for converting the set

$$X = \prod_{i=1}^{n} X_i$$

into a metric space; that is, for defining a distance function on X.

THEOREM 2.3 Let metric spaces $(X_1, d_1), (X_2, d_2), \ldots, (X_n, d_n)$ be given and set

$$X = \prod_{i=1}^{n} X_i.$$

For each pair of points $x = (x_1, x_2, \ldots, x_n)$, $y = (y_1, y_2, \ldots, y_n) \in X$, let $d:X \times X \to R$ be the function defined by the correspondence

$$d(x, y) = \underset{1 \leqq i \leqq n}{\text{maximum}} \{d_i(x_i, y_i)\}.$$

Then (X, d) is a metric space.

Proof. With x and y as above, $d_i(x_i, y_i) \geqq 0$ for $1 \leqq i \leqq n$, and therefore $d(x, y) \geqq 0$. If $d(x, y) = 0$, then $d_i(x_i, y_i) = 0$ for $1 \leqq i \leqq n$ and therefore $x_i = y_i$ for each i. Consequently, $x = y$. Conversely, if $x = y$, then $d_i(x_i, y_i) = 0$ for each i, and $d(x, y) = 0$. Since $d_i(x_i, y_i) = d_i(y_i, x_i)$ for $1 \leqq i \leqq n$, $d(x, y) = d(y, x)$. Finally, let $z = (z_1, z_2, \ldots, z_n) \in X$. Let j and k be integers such that $d(x, y) = d_j(x_j, y_j)$ and $d(y, z) = d_k(y_k, z_k)$. Thus, for $1 \leqq i \leqq n$, $d_i(x_i, y_i) \leqq d_j(x_j, y_j)$, $d_i(y_i, z_i) \leqq d_k(y_k, z_k)$, and

$$d_i(x_i, z_i) \leqq d_i(x_i, y_i) + d_i(y_i, z_i) \leqq d_j(x_j, y_j) + d_k(y_k, z_k) = d(x, y) + d(y, z).$$

Therefore $d(x, z) = \underset{1 \leqq i \leqq n}{\text{maximum}} \{d_i(x_i, z_i)\} \leqq d(x, y) + d(y, z)$.

As an immediate application of this theorem, we have:

COROLLARY 2.4 (R^n, d) is a metric space, where $d: R^n \times R^n \to R$ is the function defined by the correspondence

$$d((x_1, x_2, \ldots, x_n), (y_1, y_2, \ldots, y_n)) = \underset{1 \leqq i \leqq n}{\text{maximum}} \{|x_i - y_i|\}, \quad (x_1, x_2, \ldots, x_n), (y_1, y_2, \ldots, y_n) \in R^n.$$

It is interesting to compare the metric space (R^2, d) that we obtain in the above manner with what might be considered a more natural model of the coordinate plane. In (R^2, d) as defined above, the distance from the point $(1, 2)$ to the point $(3, 1)$ is 2, since maximum $\{|1 - 3|, |2 - 1|\} = 2$. The distance function d' used in analytical geometry would yield

$$d'((1, 2), (3, 1)) = \sqrt{(1 - 3)^2 + (2 - 1)^2} = \sqrt{5}.$$

If, for each pair of points $(x_1, x_2), (y_1, y_2) \in R^2$ we define

$$d'((x_1, x_2), (y_1, y_2)) = \sqrt{(x_1 - y_1)^2 + (x_2 - y_2)^2},$$

then we are constructing a new metric space (R^2, d'), (provided, of course, that d' is a distance function), which must be distinguished from the metric space (R^2, d) where

$$d((x_1, x_2), (y_1, y_2)) = \text{maximum } \{|x_1 - y_1|, |x_2 - y_2|\}.$$

For example, in (R^2, d) the set M of points x such that $d(x, a) \leqq 1$ for a fixed point $a \in R^2$ is a square of width 2 whose center is at a and whose sides are parallel to the coordinate axes, whereas in (R^2, d') the set of points x such that $d'(x, a) \leqq 1$ for a fixed point $a \in R^2$ is a circular disc whose center is a and whose radius is 1 (see Figure 4).

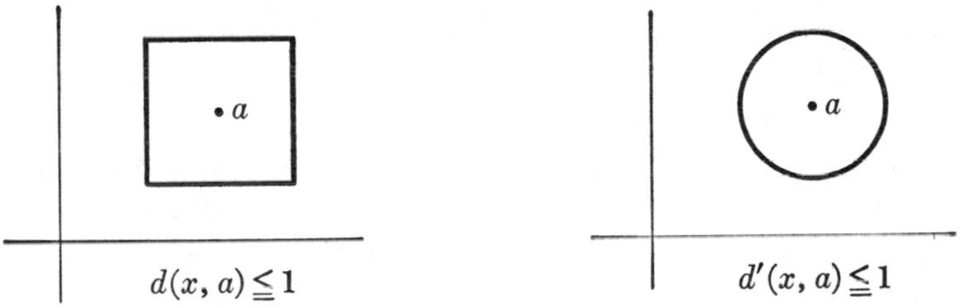

Figure 4

The formula used to define the function d' may be generalized to yield a distance function for R^n, often referred to as the *Euclidean* distance function.

THEOREM 2.5 (R^n, d') is a metric space, where d' is the function defined by the correspondence

$$d'(x, y) = \sqrt{\sum_{i=1}^{n} (x_i - y_i)^2},$$

for $x = (x_1, x_2, \ldots, x_n)$, $y = (y_1, y_2, \ldots, y_n) \in R^n$.

The proof of this theorem will be found in Section 8.

The fact that we have metric spaces (R^n, d) and (R^n, d'), with d and d' defined as above, serves to emphasize the fact that a metric space consists of two objects, a set and a distance function. Two metric spaces may be distinct even though the underlying sets of points of the two spaces are the same.

EXERCISES

1. Let (X, d) be a metric space. Let k be a positive real number and set $d_k(x, y) = k \cdot d(x, y)$. Prove that (X, d_k) is a metric space.
2. Prove that (R^n, d'') is a metric space, where the function d'' is defined by the correspondence

$$d''(x, y) = \sum_{i=1}^{n} |x_i - y_i|,$$

for $x = (x_1, x_2, \ldots, x_n)$, $y = (y_1, y_2, \ldots, y_n) \in R^n$. In (R^2, d'') determine the shape and position of the set of points x such that $d''(x, a) \leqq 1$ for a point $a \in R^2$.

3. Let d be the distance function defined on R^n by using Theorem 2.3, let d' be the Euclidean distance function, and let d'' be the distance function defined in Problem 2 above. Prove that for each pair of points $x, y \in R^n$,

$$d(x, y) \leqq d'(x, y) \leqq \sqrt{n}\, d(x, y),$$
$$d(x, y) \leqq d''(x, y) \leqq n \cdot d(x, y).$$

4. Let X be the set of all continuous functions $f:[a, b] \to R$. For $f, g \in X$, define

$$d(f, g) = \int_a^b |f(t) - g(t)|\, dt.$$

Using appropriate theorems from Calculus, prove that (X, d) is a metric space.

5. Let $S \subset R$. A function $f:S \to R$ is called *bounded* if there is a real number K such that $|f(x)| \leqq K$, $x \in S$ (or equivalently, $f(S) \subset [-K, K]$). Let X' be the set of all bounded functions $f:[a, b] \to R$. For $f, g \in X'$ define

$$d'(f, g) = \text{l.u.b.} \cup_{x \in [a,b]} \{|f(x) - g(x)|\},$$

(l.u.b. is an abbreviation of *least upper bound*, see Definition 5.5 of this chapter). Prove that (X', d') is a metric space.

6. Let $f, g:[a, b] \to R$ be two functions that are both continuous and

bounded. Compare $d(f, g)$ and $d'(f, g)$, where d and d' are defined as in Problems 4 and 5 respectively.

7. Let X be a set. For $x, y \in X$ define the function d by

$$d(x, x) = 0,$$

and

$$d(x, y) = 1,$$

if $x \neq y$. Prove that (X, d) is a metric space.

8. Let Z be the set of integers. Let p be a positive prime integer. Given distinct integers m, n there is a unique integer $t = t(m, n)$ such that $m - n = p^t \cdot k$, where k is an integer not divisible by p. Define a function $d: Z \times Z \to R$ by the correspondence $d(m, m) = 0$ and

$$d(m, n) = \frac{1}{p^t}$$

for $m \neq n$. Prove that (Z, d) is a metric space. [*Hint:* for $a, b, c \in Z$, $t(a, c) \geq$ minimum $\{t(a, b), t(b, c)\}$]. Let $p = 3$. What is the set of elements $x \in Z$ such that $d(0, x) < 1$? What is the set of elements $x \in Z$ such that $d(0, x) < \frac{1}{3}$?

3 CONTINUITY

In calculus, the first occurrence of the word "continuity" is with reference to a function $f: R \to R$, R the set of real numbers. To decide which condition or conditions this function must satisfy for us to say, "the function f is continuous at a point $a \in R$," we try to decide upon a precise formulation of the statement "a number $f(x)$ will be close to the number $f(a)$ whenever the number x is close to a." Having defined a distance function for the real numbers R, we have a quantitative measure of the degree of closeness of two numbers. But how close must $f(x)$ be to $f(a)$? Instead of specifying some particular degree of closeness of $f(x)$ to $f(a)$, let us think, rather, of requiring that no matter what choice is made for the degree of closeness of $f(x)$ to $f(a)$, it can be so arranged that this degree of closeness is achieved. By the phrase "arrange matters" we mean that we can find a corresponding degree of closeness so that whenever x is within this corresponding degree of closeness to a, then $f(x)$ is within the prescribed degree of closeness to $f(a)$. We have now arrived at the following formulation, "the function $f: R \to R$ is continuous at the number $a \in R$, if given a prescribed degree of closeness, $f(x)$ will be within this prescribed degree of closeness to $f(a)$, whenever x is within some corresponding degree of closeness to a." To put this statement in its final form, we shall substitute for "a prescribed degree of closeness" the symbol "ε," and for the phrase "some corresponding degree of closeness" the symbol "δ," and use the distance function to measure the degree of closeness.

DEFINITION 3.1 Let $f: R \to R$. The function f is said to be *continuous at the point* $a \in R$, if given $\varepsilon > 0$, there is a $\delta > 0$, such that

$$|f(x) - f(a)| < \varepsilon,$$

whenever

$$|x - a| < \delta.$$

The function f is said to be *continuous* if it is continuous at each point of R.

Because we initially formulated the definition of continuity in terms of the phrase "degree of closeness," we may easily devise a definition of "continuity" applicable to metric spaces in general, since we need only use the distance functions of these metric spaces to measure "degree of closeness."

DEFINITION 3.2 Let (X, d) and (Y, d') be metric spaces, and let $a \in X$. A function $f: X \to Y$ is said to be *continuous at the point* $a \in X$ if given $\varepsilon > 0$, there is a $\delta > 0$, such that

$$d'(f(x), f(a)) < \varepsilon$$

whenever $x \in X$ and

$$d(x, a) < \delta.$$

The function $f:X \to Y$ is said to be *continuous* if it is continuous at each point of X.

Definitions, such as those given above, are created to serve two purposes. First of all, they are abbreviations. Thus, the statement that begins, "given $\varepsilon > 0$, there is . . . ," is replaced by the shorter statement, "$f:X \to Y$ is continuous at the point $a \in X$." Second, these definitions are attempts to formulate precise characterizations of what we feel are significant properties; in this case, the property of being continuous at a point. We have tried to indicate in the discussion preceding these definitions that they do provide a precise characterization of our intuitive, and perhaps vague, concept of continuity. There are, in a certain sense, tests that we may apply to see whether or not they do so. As an illustration, there are certain functions that we "know" are "continuous," that is, we are sure that they possess this property we are trying to characterize. If it should turn out that a function we "know" to be "continuous" is not continuous in accordance with these definitions, then, although these definitions may be precise, they would not furnish a precise characterization of the property we have in mind when we say a function is "continuous." This type of testing of a definition thus takes the form of proving theorems to the effect that certain functions are continuous. For example:

THEOREM 3.3 Let (X, d) and (Y, d') be metric spaces. Let $f:X \to Y$ be a constant function, then f is continuous.

Proof. Let a point $a \in X$ and $\varepsilon > 0$ be given. Choose any $\delta > 0$, say $\delta = 1$. Then whenever $d(x, a) < \delta$, we have $d'(f(x), f(a)) = 0 < \varepsilon$.

THEOREM 3.4 Let (X, d) be a metric space. Then the identity function $i:X \to X$ is continuous.

Proof. Suppose $a \in X$. Let $\varepsilon > 0$ be given. Choose $\delta = \varepsilon$, then whenever $d(x, a) < \delta$ we have $d(i(x), i(a)) = d(x, a) < \varepsilon$.

Note that in the above proof we could have equally well chosen δ to be any positive number, provided only that $\delta \leqq \varepsilon$, and the proof would still be valid. The choice of δ need not be a very efficient choice; all that is required is that it "do the job."

There is one situation we shall have to consider for which the notation $f:X \to Y$ that we have adopted for a function from a metric space (X, d) into a metric space (Y, d') is ambiguous. Consider metric spaces (X, d) and (X, d') with the same underlying set. If we simply write $f:X \to X$ for a function, it is impossible to tell which metric space is denoted by the first occurrence of X and which by the second. For this reason, when considering one set X with two different distance functions, we shall write $f:(X, d) \to (X, d')$ if we intend to think of $f:X \to X$ as a function from the metric space (X, d) into the metric space (X, d'). As an illustration, we shall prove:

THEOREM 3.5 Let $i:R^n \to R^n$ be the identity function. Then

$$i:(R^n, d) \to (R^n, d')$$

and

$$i:(R^n, d') \to (R^n, d)$$

are continuous, where the distance function d is the maximum distance between corresponding coordinates (as defined in Section 2) and d' is the Euclidean distance.

Proof. Let $a = (a_1, a_2, \ldots, a_n) \in R^n$. We shall first prove that $i:(R^n, d) \to (R^n, d')$ is continuous. Let $\varepsilon > 0$ be given. Choose $\delta = \varepsilon/\sqrt{n}$. Suppose $x = (x_1, x_2, \ldots, x_n)$ is such that $d(x, a) < \delta$; that is, $\underset{1 \leqq i \leqq n}{\text{maximum}} \{|a_i - x_i|\} < \delta$. Then

$$d'(x, a) = \sqrt{\sum_{i=1}^{n} (a_i - x_i)^2} < \sqrt{n\delta^2} = \sqrt{\varepsilon^2} = \varepsilon.$$

Therefore, given $\varepsilon > 0$, there is a $\delta > 0$ such that $d'(i(x), i(a)) < \varepsilon$ whenever $d(x, a) < \delta$.

We now prove that $i:(R^n, d') \to (R^n, d)$ is continuous. Let $\varepsilon > 0$ be given. Choose $\delta = \varepsilon$. Suppose that $x = (x_1, x_2, \ldots, x_n)$ is such that $d'(x, a) < \delta$. Then

$$\sum_{i=1}^{n} (a_i - x_i)^2 < \delta^2$$

and therefore for each i, $(a_i - x_i)^2 < \delta^2$, or $|a_i - x_i| < \delta = \varepsilon$. Consequently, $d(x, a) < \varepsilon$. Thus, given $\varepsilon > 0$, there is a $\delta > 0$, such that $d(i(x), i(a)) < \varepsilon$ whenever $d'(x, a) < \delta$.

One of the most important elementary theorems about continuous functions is the statement that the composition of two continuous functions is again a continuous function.

THEOREM 3.6 Let (X, d), (Y, d'), (Z, d'') be metric spaces. Let $f\colon X \to Y$ be continuous at the point $a \in X$ and let $g\colon Y \to Z$ be continuous at the point $f(a) \in Y$. Then $gf\colon X \to Z$ is continuous at the point $a \in X$.

Proof. Let $\varepsilon > 0$ be given. We must find a $\delta > 0$ such that whenever $x \in X$ and $d(x, a) < \delta$, then $d''(g(f(x)), g(f(a))) < \varepsilon$. Since g is continuous at $f(a)$, there is an $\eta > 0$, such that whenever $y \in Y$ and $d'(y, f(a)) < \eta$, then $d''(g(y), g(f(a))) < \varepsilon$. Using the fact that f is continuous at a, we know that given $\eta > 0$, there is a $\delta > 0$, such that $x \in X$ and $d(x, a) < \delta$ imply that $d'(f(x), f(a)) < \eta$ and hence $d''(g(f(x)), g(f(a))) < \varepsilon$.

COROLLARY 3.7 Let (X, d), (Y, d'), (Z, d'') be metric spaces. Let $f\colon X \to Y$ and $g\colon Y \to Z$ be continuous. Then $gf\colon X \to Z$ is continuous.

EXERCISES

1. Let X be the set of continuous functions $f\colon [a, b] \to R$. Let d^* be the distance function on X defined by

$$d^*(f, g) = \int_a^b |f(t) - g(t)|\, dt,$$

for $f, g \in X$. For each $f \in X$, set

$$I(f) = \int_a^b f(t)\, dt.$$

Prove that the function $I\colon (X, d^*) \to (R, d)$ is continuous.

2. Let (X_i, d_i), (Y_i, d_i'), $i = 1, \ldots, n$ be metric spaces. Let $f_i: X_i \to Y_i$, $i = 1, \ldots, n$ be continuous functions. Let

$$X = \prod_{i=1}^{n} X_i \quad \text{and} \quad Y = \prod_{i=1}^{n} Y_i$$

and convert X and Y into metric spaces in the standard manner. Define the function $F: X \to Y$ by

$$F(x_1, x_2, \ldots, x_n) = (f_1(x_1), f_2(x_2), \ldots, f_n(x_n)).$$

Prove that F is continuous.

3. Define the function $f: R^2 \to R$ by $f(x_1, x_2) = x_1 + x_2$. Prove that f is continuous, where the distance function on R^2 is either d or d'.
4. Define functions g, h, k, m as follows: $g: R^2 \to R^2 \times R^2$ by $g(x, y) = ((x, y), (x, y))$; $h: R^2 \times R^2 \to R \times R$ by $h((a, b), (c, d)) = (a + b, c - d)$; $k: R \times R \to R \times R$ by $k(u, v) = (u^2, v^2)$; $m: R \times R \to R$ by $m(x, y) = \frac{1}{4}(x - y)$. Prove that all these functions are continuous and that $xy = mkhg(x, y)$.

4 OPEN BALLS AND NEIGHBORHOODS

In the definition of continuity of a function f at a point a in a metric space (X, d), we are concerned with how f transforms those points $x \in X$ such that $d(x, a) < \delta$. If we give a name to this particular collection of points in X we shall be able to cast the definition of continuity in a more compact form.

Definition 4.1 Let (X, d) be a metric space. Let $a \in X$ and $\delta > 0$ be given. The subset of X consisting of those points $x \in X$ such that $d(a, x) < \delta$ is called the *open ball about a of radius δ* and is denoted by

$$B(a; \delta).$$

Thus, $x \in B(a; \delta)$ if and only if $x \in X$ and $d(x, a) < \delta$. Similarly, if (Y, d') is another metric space and $f:X \to Y$, we have $y \in B(f(a); \varepsilon)$ if and only if $y \in Y$ and $d'(y, f(a)) < \varepsilon$. Thus:

THEOREM 4.2 A function $f:(X, d) \to (Y, d')$ is continuous at a point $a \in X$ if and only if given $\varepsilon > 0$ there is a $\delta > 0$ such that

$$f(B(a; \delta)) \subset B(f(a); \varepsilon).$$

For a function $f:X \to Y$ we have $f(U) \subset V$ if and only if $U \subset f^{-1}(V)$, where U and V are subsets of X and Y respectively. Therefore:

THEOREM 4.3 A function $f:(X, d) \to (Y, d')$ is continuous at a point $a \in X$ if and only if given $\varepsilon > 0$ there is a $\delta > 0$ such that

$$B(a; \delta) \subset f^{-1}(B(f(a); \varepsilon)).$$

Given a point a in a metric space (X, d), the subset $B(a; \delta)$ of X, for each $\delta > 0$, is an example of the type of subset of X that is called a neighborhood of a.

DEFINITION 4.4 Let (X, d) be a metric space and $a \in X$. A subset N of X is called a *neighborhood of* a if there is a $\delta > 0$ such that

$$B(a; \delta) \subset N.$$

The collection $\mathfrak{N}_a$ of all neighborhoods of a point $a \in X$ is called a *complete system of neighborhoods* of the point a.

A neighborhood of a point $a \in X$ may be thought of as containing all the points of X that are sufficiently close to a or as "enclosing" a by virtue of the fact that it contains some open ball about a. In particular, for each $\delta > 0$, $B(a; \delta)$ is a neighborhood of a. These open balls have the property that they are neighborhoods of each of their points.

Lemma 4.5 Let (X, d) be a metric space and $a \in X$. For each $\delta > 0$, the open ball $B(a; \delta)$ is a neighborhood of each of its points.

Proof. Let $b \in B(a; \delta)$. In order to show that $B(a; \delta)$ is a neighborhood of b we must show that there is an $\eta > 0$ such that $B(b; \eta) \subset B(a; \delta)$. Since $b \in B(a; \delta)$, $d(a, b) < \delta$. Choose $\eta < \delta - d(a, b)$. If $x \in B(b; \eta)$ then

$$d(a, x) \leqq d(a, b) + d(b, x) < d(a, b) + \eta < d(a, b) + \delta - d(a, b) = \delta,$$

and therefore $x \in B(a; \delta)$. Thus $B(b; \eta) \subset B(a; \delta)$ and $B(a; \delta)$ is a neighborhood of b.

We may describe this proof pictorially. We have started with an open ball $B(a; \delta)$ about a. We choose a point $b \in B(a; \delta)$. Then the minimum distance from b to points not in $B(a; \delta)$ is at least $\delta - d(a, b)$, as indicated in Figure 5, so that a ball about b

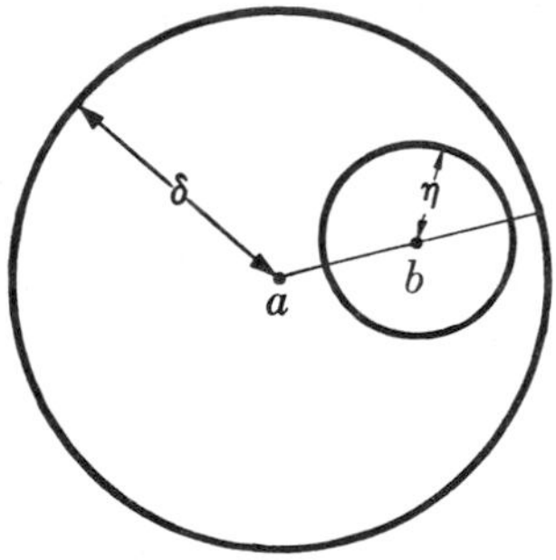

Figure 5

of radius $\eta < \delta - d(a, b)$ is contained in $B(a; \delta)$.

The complete system of neighborhoods of a point may be used to characterize continuity of a function at a point.

Theorem 4.6 Let $f: (X, d) \to (Y, d')$. f is continuous at a point $a \in X$ if and only if for each neighborhood M of $f(a)$ there is a corresponding neighborhood N of a, such that

$$f(N) \subset M,$$

or equivalently,

$$N \subset f^{-1}(M).$$

Proof. First suppose that f is continuous at the point $a \in X$. We must show that, given a neighborhood M of $f(a)$, we can find a neighborhood N of a such that $f(N) \subset M$. Since M is a neighborhood of $f(a)$, there is an $\varepsilon > 0$ such that $B(f(a); \varepsilon) \subset M$. Since f is continuous at a, there is a $\delta > 0$ such that $f(B(a; \delta)) \subset B(f(a); \varepsilon)$. But $N = B(a; \delta)$ is a neighborhood of a, therefore

$$f(N) = f(B(a; \delta)) \subset B(f(a); \varepsilon) \subset M.$$

Conversely, suppose that f satisfies the property that for each neighborhood M of $f(a)$, there is a corresponding neighborhood N of a, such that $f(N) \subset M$. Let $\varepsilon > 0$ be given. To prove that f is continuous at a we must show that there is a $\delta > 0$ such that

$$f(B(a; \delta)) \subset B(f(a); \varepsilon).$$

But $B(f(a); \varepsilon) = M$ is a neighborhood of $f(a)$ and therefore there is a neighborhood N of a such that $f(N) \subset M$. Since N is a neighborhood of a, there is a $\delta > 0$ such that $B(a; \delta) \subset N$. Therefore

$$f(B(a; \delta)) \subset f(N) \subset M = B(f(a); \varepsilon).$$

The proof of the first part of the above theorem may be represented pictorially by considering an arbitrary neighborhood M of $f(a)$ (as indicated in Figure 6). Since M is a neigh-

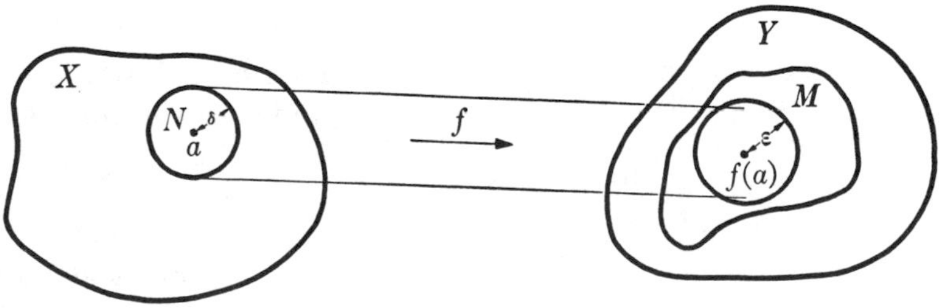

Figure 6

borhood of $f(a)$, it contains an open ball $B(f(a); \varepsilon)$ for some $\varepsilon > 0$. Since f is continuous at a, for some $\delta > 0$ the neighborhood $N = B(a; \delta)$ is carried into M by f. Similarly, the proof of the second part of the theorem may be depicted by Figure 7. We start with a neighborhood $M = B(f(a); \varepsilon)$ of $f(a)$. The assumed

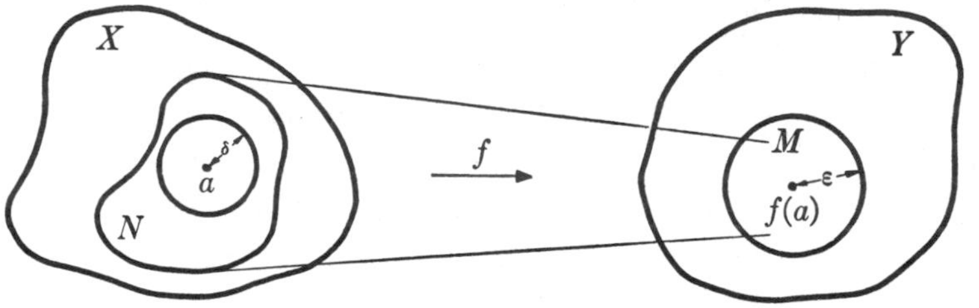

Figure 7

property of f allows us to assert that there is a neighborhood N of a that is carried into M by f. Since N is a neighborhood of a we have an open ball $B(a; \delta)$ contained in N, which must also be carried into M.

If N is a neighborhood of a point a in a metric space (X, d) and N' is a subset of X that contains N, then N' contains the same open ball about a that N does and therefore N' is also a neighborhood of a. Thus, the previous theorem becomes:

THEOREM 4.7 Let $f: (X, d) \to (Y, d')$. f is continuous at a point $a \in X$ if and only if for each neighborhood M of $f(a)$, $f^{-1}(M)$ is a neighborhood of a.

The collections of neighborhoods of points in a metric space possess five properties that will be of significance in the next chapter.

THEOREM 4.8 Let (X, d) be a metric space.

$N1$. For each point $a \in X$, there exists at least one neighborhood of a.

$N2$. For each point $a \in X$ and each neighborhood N of a, $a \in N$.

*N*3. For each point $a \in X$, if N is a neighborhood of a and $N' \supset N$, then N' is a neighborhood of a.

*N*4. For each point $a \in X$ and each pair N, M of neighborhoods of a, $N \cap M$ is also a neighborhood of a.

*N*5. For each point $a \in X$ and each neighborhood N of a, there exists a neighborhood O of a such that $O \subset N$ and O is a neighborhood of each of its points.

Proof. For $a \in X$, X is a neighborhood of a, thus $N1$ is true. $N2$ is trivial and $N3$ has already been discussed. To prove $N4$, let N and M be neighborhoods of a. Then N and M contain open balls $B(a; \delta_1)$ and $B(a; \delta_2)$ respectively and therefore $N \cap M$ contains the open ball $B(a; \delta)$, where $\delta = \text{minimum } \{\delta_1, \delta_2\}$. To prove $N5$, let N be a neighborhood of a. Then N contains an open ball $B(a; \delta)$ and by Lemma 4.5, $O = B(a; \delta)$ is a neighborhood of each of its points.

For a given point a in a metric space X, the collection of open balls with center a has been used to generate the complete system of neighborhoods at a, in the sense that the neighborhoods of a are precisely those subsets of X which contain one of these open balls.

DEFINITION 4.9 Let a be a point in a metric space X. A collection $\mathcal{B}_a$ of neighborhoods of a is called a *basis for the neighborhood system at a* if every neighborhood N of a contains some element B of $\mathcal{B}_a$.

As an example, if a is a point on the real line R, a basis for the neighborhood system at a is the collection of open intervals containing a.

EXERCISES

1. Let (X, d) be a metric space such that $d(x, y) = 1$ whenever $x \neq y$. Let $a \in X$. Prove that $\{a\}$ is a neighborhood of a and constitutes a basis for the system of neighborhoods at a. Prove that every subset of X is a neighborhood of each of its points.

2. Let $a \in R$ and $f:R \to R$ be defined by $f(x) = 0$ for $x \leqq a$, $f(x) = 1$ for $x > a$. Prove that f is not continuous at a, but is continuous at all other points.
3. Let $f:X \to Y$ be a function from a metric space X into a metric space Y. Let $a \in X$ and let $\mathcal{B}_{f(a)}$ be a basis for the neighborhood system at $f(a)$. Prove that f is continuous at a if and only if for each $N \in \mathcal{B}_{f(a)}$, $f^{-1}(N)$ is a neighborhood of a.
4. Let a be a point on the real line R. Prove that each of the following collections of subsets of R constitute a basis for the system of neighborhoods at a:

   i) All closed intervals of the form $[a - \varepsilon, a + \varepsilon]$, $\varepsilon > 0$;
   ii) All open balls $B(a; \varepsilon)$, ε a positive rational number;
   iii) All open balls $B\left(a; \frac{1}{n}\right)$, n a positive integer;
   iv) All open balls $B\left(a; \frac{1}{n}\right)$, n a positive integer larger than some fixed integer k.

   Show that no finite collection of subsets of R can be a basis for the system of neighborhoods at a.
5. Let a be a point in a metric space X. Let N be the set of positive integers. Prove that there is a collection $\{B_n\}_{n \in N}$ of neighborhoods of a which constitutes a basis for the system of neighborhoods at a.
6. Let a and b be distinct points of a metric space X. Prove that there are neighborhoods N_a and N_b of a and b respectively such that $N_a \cap N_b = \emptyset$.
7. Let $(X_1, d_1), (X_2, d_2), \ldots, (X_n, d_n)$ be metric spaces and convert

$$X = \prod_{i=1}^{n} X_i$$

into a metric space (X, d) in the standard manner. Prove that an open ball in (X, d) is the product of open balls from $X_1, X_2, \ldots, X_n$ respectively. Let $a_i \in X_i$, $i = 1, 2, \ldots, n$, and let $\mathcal{B}_{a_i}$ be a basis for the neighborhood system at a_i. Let $\mathcal{B}_a$ be the collection of all sets of the form $B_1 \times B_2 \times \ldots \times B_n$, $B_i \in \mathcal{B}_{a_i}$. Prove that $\mathcal{B}_a$ is a basis for the neighborhood system at $a = (a_1, a_2, \ldots, a_n) \in X$. Let $p_i:X \to X_i$, $i = 1, 2, \ldots, n$, be the projection that maps $p_i(a) = a_i$. Prove that each p_i is continuous. Let Y be a metric space and $f:Y \to X$ a function. Prove that f is continuous if and only if each of the n functions $p_i f$ is continuous.

8. Let R be the real numbers and $f: R \to R$ a continuous function. Suppose that for some number $a \in R$, $f(a) > 0$. Prove that there is a positive number k and a closed interval $F = [a - \delta, a + \delta]$ for some $\delta > 0$ such that $f(x) \geqq k$ for $x \in F$.

5 LIMITS

The concept of limit of a sequence of real numbers may be generalized to an arbitrary metric space. First, let us recall the appropriate definition in the real line.

DEFINITION 5.1 Let $a_1, a_2, \ldots$ be a sequence of real numbers. A real number a is said to be the *limit of the sequence* $a_1, a_2, \ldots$ if, given $\varepsilon > 0$, there is a positive integer N such that, whenever $n > N$, $|a - a_n| < \varepsilon$. In this event we shall also say that the sequence $a_1, a_2, \ldots$ *converges to* a and write $\lim_n a_n = a$.

Interpreting ε as an "arbitrary degree of closeness" and N as "sufficiently far out in the sequence," we see that we have defined $\lim_n a_n = a$ in the event that a_n may be made arbitrarily close to a by requiring that a_n be sufficiently far out in the sequence.

Now, suppose that we have a metric space (X, d) and a sequence $a_1, a_2, \ldots$ of points of X. Given a point $a \in X$ we measure the distance from a to the successive points of the sequence, by the sequence of real numbers $d(a, a_1), d(a, a_2), \ldots$. It is natural to say that the limit of the sequence $a_1, a_2, \ldots$ of points of X is the point a if the limit of the sequence of real numbers $d(a, a_1), d(a, a_2), \ldots$ is the real number 0.

DEFINITION 5.2 Let (X, d) be a metric space. Let $a_1, a_2, \ldots$ be a sequence of points of X. A point $a \in X$ is said to be the *limit of the sequence* $a_1, a_2, \ldots$ if $\lim_n d(a, a_n) = 0$. Again, in this event, we shall say that the sequence

$a_1, a_2, \ldots$ *converges to* a and write $\lim_n a_n = a$.

COROLLARY 5.3 Let (X, d) be a metric space and $a_1, a_2, \ldots$ be a sequence of points of X. Then $\lim_n a_n = a$ for a point $a \in X$ if and only if for each neighborhood V of a there is an integer N such that $a_n \in V$ whenever $n > N$.

Proof. Let V be a neighborhood of a. For some $\varepsilon > 0$, $a \in B(a; \varepsilon) \subset V$. Thus if $\lim_n a_n = a$ there is an integer N such that whenever $n > N$, $d(a, a_n) < \varepsilon$ and hence $a_n \in V$. Conversely, given $\varepsilon > 0$, $B(a; \varepsilon)$ is a neighborhood of a. If there is an integer N such that for $n > N$, $a_n \in B(a; \varepsilon)$, then $d(a, a_n) < \varepsilon$ and $\lim_n a_n = a$.

If S is a set of infinite points, and there is at most a finite number of elements of S for which a certain statement is false, then the statement is said to be true for *almost all* the elements of S. Thus $\lim_n a_n = a$ if for each neighborhood V of a almost all the points a_n are in V.

Continuity may be characterized in terms of limits of sequences in accordance with the following theorem.

THEOREM 5.4 Let (X, d), (Y, d') be metric spaces. A function $f: X \to Y$ is continuous at a point $a \in X$ if and only if, whenever $\lim_n a_n = a$ for a sequence $a_1, a_2, \ldots$ of points of X, $\lim_n f(a_n) = f(a)$.

Proof. Suppose f is continuous at a and $\lim_n a_n = a$. Let V be a neighborhood of $f(a)$. Then $f^{-1}(V)$ is a neighborhood of a, so by Corollary 5.3 there is an integer N such that $a_n \in f^{-1}(V)$ whenever $n > N$. Consequently, $f(a_n) \in V$ whenever $n > N$. Thus, for each neighborhood V of $f(a)$ there is an integer N such that $f(a_n) \in V$ whenever $n > N$ and again, applying Corollary 5.3, $\lim_n f(a_n) = f(a)$.

To prove the "if" part of this theorem, we shall prove that if f is not continuous at a, then there is at least one sequence $a_1, a_2, \ldots$ of points of X, such that $\lim_n a_n = a$, but $\lim_n f(a_n) = f(a)$ is false. Since f is not continuous at a, there is a neighborhood V of $f(a)$ such that for each neighborhood U of a, $f(U) \not\subset V$. In particular, for each neighborhood $B\left(a; \frac{1}{n}\right)$, $n = 1, 2, \ldots$ $f\left(B\left(a; \frac{1}{n}\right)\right) \not\subset V$. Thus, for each positive integer n, there is a point a_n with $a_n \in B\left(a; \frac{1}{n}\right)$ and $f(a_n) \notin V$. Now $d(a, a_n) < \frac{1}{n}$ and therefore $\lim_n a_n = a$, whereas, $\lim_n f(a_n) = f(a)$ is impossible, since $f(a_n) \notin V$ for all n.

If $\lim_n a_n = a$, we can write $\lim_n f(a_n) = f(a)$ as $\lim_n f(a_n) = f(\lim_n a_n)$. We may therefore describe a continuous function as one that commutes with the operation of taking limits. It is worth noting that in proving f is continuous whenever f commutes with the operation of taking limits we have used the fact that the sequence of neighborhoods $B\left(a; \frac{1}{n}\right)$, n a positive integer, constitutes a basis for the neighborhood system at a.

In order to introduce the concept of distance from a point to a subset we shall recall some facts about the real number system.

DEFINITION 5.5 Let A be a set of real numbers. A number b is called an *upper bound* of A if $x \leqq b$ for each $x \in A$. A number c is called a *lower bound* of A if $c \leqq x$ for each $x \in A$. If A has both an upper and lower bound A is said to be *bounded*.

An upper bound b^* of A is called a *least upper bound* (abbreviated l.u.b.) of A if for each upper bound b of A, $b^* \leqq b$. A lower bound c^* of A is called a *greatest lower bound* (abbreviated g.l.b.) of A if for each lower bound c of A, $c \leqq c^*$.

Not every set of real numbers has an upper bound. One of the properties of the real number system, usually referred to as the *completeness postulate*, is that a non-empty set A of real numbers which has an upper bound necessarily has a l.u.b. Given a non-empty set B of real numbers which has a lower bound, the set of negatives of elements of B has an upper bound, hence a l.u.b. whose negative is a g.l.b. of B. Thus it follows that every non-empty set B of real numbers which has a lower bound has a g.l.b.

The greatest lower bound of a set A of real numbers may or may not be an element of A. For example, 0 is a g.l.b. of $[0, 1]$ and $0 \in [0, 1]$, whereas 0 is also a g.l.b. of $(0, 1)$ but $0 \notin (0, 1)$. In any event, the g.l.b. of a set of real numbers must be arbitrarily close to that set.

LEMMA 5.6 Let b be a greatest lower bound of the non-empty subset A. Then, for each $\varepsilon > 0$, there is an element $x \in A$ such that

$$x - b < \varepsilon.$$

Proof. Suppose there were an $\varepsilon > 0$ such that $x - b \geqq \varepsilon$ for each $x \in A$. Then $b + \varepsilon \leqq x$ for each $x \in A$ and $b + \varepsilon$ would be a lower bound of A. Since b is a g.l.b. of A, we obtain the contradiction $b + \varepsilon \leqq b$.

COROLLARY 5.7 Let b be a greatest lower bound of the non-empty subset A of real numbers. Then there is a sequence $a_1, a_2, \ldots$ of real numbers such that $a_n \in A$ for each n and $\lim_n a_n = b$.

Proof. For $\varepsilon = \frac{1}{n}$ we obtain an element $a_n \in A$ such that $a_n - b < \frac{1}{n}$. Since b is a lower bound of A, $0 \leqq a_n - b$. Therefore $\lim_n a_n = b$.

DEFINITION 5.8 Let (X, d) be a metric space. Let $a \in X$ and let A be a non-empty subset of X. The greatest lower bound of the set of numbers of the form $d(a, x)$ for $x \in A$ is called the *distance between a and A* and is denoted by $d(a, A)$.

From Corollary 5.7 we obtain:

COROLLARY 5.9 Let (X, d) be a metric space, $a \in X$, and A a non-empty subset of X. Then there is a sequence $a_1, a_2, \ldots$ of points of A such that $\lim_n d(a, a_n) = d(a, A)$.

EXERCISES

1. Let $X_1, X_2, \ldots, X_k$ be metric spaces and convert $X = \prod_{i=1}^{k} X_i$ into a metric space in the standard manner. Each of the points $a_1, a_2, \ldots$ of a sequence of points of X has k coordinates; that is $a_n = (a_1^n, a_2^n, \ldots, a_k^n) \in X$, $n = 1, 2, \ldots$. Let $c = (c_1, c_2, \ldots, c_k) \in X$. Prove that $\lim_n a_n = c$ if and only if $\lim_n a_i^n = c_i$, $i = 1, 2, \ldots, k$.
2. In each of the three metric spaces (R^k, d), (R^k, d'), (R^k, d'') considered in Section 2, prove that limits of sequences are the same.
3. Prove that a subsequence of a convergent sequence is convergent and converges to the same limit as the original sequence.
4. A sequence of real numbers $a_1, a_2, \ldots$ is called *monotone non-decreasing* if $a_i \leqq a_{i+1}$ for each i and called *monotone non-increasing* if $a_i \geqq a_{i+1}$ for each i. A sequence which is either monotone non-decreasing or monotone non-increasing is said to be *monotone*. The sequence is said to be *bounded above* if there is a number K such that $a_i \leqq K$ for each i and *bounded below* if there is a number M such that $a_i \geqq M$ for each i. A sequence which is both bounded above and bounded below is called *bounded*. Prove that a convergent sequence of real numbers is bounded. Prove that a monotone non-decreasing sequence of real numbers which is bounded above converges to a limit a and that a is the l.u.b. of the set $\{a_1, a_2, \ldots\}$. Similarly prove that a monotone non-increasing sequence which is bounded below converges to a limit b and that b is the g.l.b. of the set $\{a_1, a_2, \ldots\}$.
5. Let $a_1, a_2, \ldots$ be a bounded sequence of real numbers. Since each of the sets $A_n = \{a_n, a_{n+1}, \ldots\}$ is bounded we may set $v_n =$ g.l.b. A_n, $u_n =$ l.u.b. A_n. Observe that $v_n \leqq u_n$; $v_1, v_2, \ldots$ is monotone non-decreasing and bounded above; and $u_1, u_2, \ldots$ is monotone non-increasing and bounded below. Let $V = \lim_n v_n$ and $U = \lim_n u_n$. Prove that there are subsequences of $a_1, a_2, \ldots$ which converge to U and V respectively (thus a bounded sequence of real numbers has a convergent subsequence). Prove that $a_1, a_2, \ldots$ converges if and only if $U = V$.

6. Let (X, d) be a metric space and A a non-empty subset of X. For $x, y \in X$, prove that $d(x, A) \leqq d(x, y) + d(y, A)$.
7. Let A be a non-empty subset of a metric space (X, d). Define the function $f: X \rightarrow R$ by $f(x) = d(x, A)$. Prove that f is continuous.
8. Let A be a non-empty subset of a metric space (X, d) and let $x \in X$. Prove that $d(x, A) = 0$ if and only if every neighborhood of x contains a point of A.
9. Let (X, d) be a metric space. Define a distance function d^* on $X \times X$ by the method of Theorem 2.3. Prove that the function $d: (X \times X, d^*) \rightarrow (R, d)$ is continuous.

6 OPEN SETS AND CLOSED SETS

In a metric space, the open ball $B(a; \delta)$ is a neighborhood of each of its points (Lemma 4.5). The collection of subsets possessing this property plays a fundamental role in topology.

DEFINITION 6.1 A subset O of a metric space is said to be *open* if O is a neighborhood of each of its points.

Open sets may be characterized directly in terms of open balls.

THEOREM 6.2 A subset O of a metric space (X, d) is an open set if and only if it is a union of open balls.

Proof. Suppose O is open. Then for each $a \in O$, there is an open ball $B(a; \delta_a) \subset O$. Therefore $O = \bigcup_{a \in O} B(a; \delta_a)$ is a union of open balls. Conversely, if O is a union of open balls, then using the centers of these balls as the elements of an indexing set we can write $O = \bigcup_{a \in I} B(a; \delta_a)$. If $x \in O$, then $x \in B(a; \delta_a)$ for some $a \in I$. $B(a; \delta_a)$ is a neighborhood of x and since $B(a; \delta_a) \subset O$, by $N3$, O is a neighborhood of x. Thus O is a neighborhood of each of its points, and by Definition 6.1, O is open.

Most of the functions considered in topology are continuous. Open sets provide a simple characterization of continuity.

Theorem 6.3 Let $f: (X, d) \to (Y, d')$. Then f is continuous if and only if for each open set O of Y, the subset $f^{-1}(O)$ is an open subset of X.

Proof. First, suppose f is continuous. Let $O \subset Y$ be open. We must show that $f^{-1}(O)$ is open; that is, $f^{-1}(O)$ is a neighborhood of each of its points. To this end, let $a \in f^{-1}(O)$, then $f(a) \in O$ and O is a neighborhood of $f(a)$. Since f is continuous at a, Theorem 4.7 may be applied, yielding $f^{-1}(O)$ is a neighborhood of a.

Conversely, suppose for each open set $O \subset Y, f^{-1}(O)$ is open. Let $a \in X$ and let M be a neighborhood of $f(a)$. Then there is an $\varepsilon > 0$ such that $B(f(a); \varepsilon) \subset M$. But $B(f(a); \varepsilon)$ is open and therefore $f^{-1}(B(f(a); \varepsilon))$ is open. Since $a \in f^{-1}(B(f(a); \varepsilon))$, this subset is a neighborhood of a. Therefore $f^{-1}(M)$ contains a neighborhood of a and f is continuous at a. Since a was arbitrary, f is continuous.

Just as the collections of neighborhoods of points in a metric space possess certain significant properties so do the collection of open sets in a metric space.

Theorem 6.4 Let (X, d) be a metric space.

$O1$. The empty set is open.

$O2$. X is open.

$O3$. If $O_1, O_2, \ldots, O_n$ are open, then $O_1 \cap O_2 \cap \ldots \cap O_n$ is open.

$O4$. If for each $\alpha \in I$, O_α is an open set, then $\bigcup_{\alpha \in I} O_\alpha$ is open.

Proof. The empty set is open, for in order for it not to be open there would have to be a point $x \in \emptyset$. Given a point $a \in X$, for any $\delta > 0$, $B(a; \delta) \subset X$ and therefore X is a neighborhood of each of its points; that is, X is open. To prove $O3$, let $a \in O_1 \cap O_2 \cap \ldots \cap O_n$, where for $i = 1, 2, \ldots, n$, O_i is open. Then each O_i is a neighborhood of a. By $N4$, the intersection of two neighborhoods of a is again a neighborhood of a, and hence by induction, the intersection of a finite number of neighborhoods of a is again a neighborhood of a. Therefore $O_1 \cap O_2 \cap \ldots \cap O_n$ is a neighborhood of each of its points. Finally, to prove $O4$, let $a \in O = \bigcup_{\alpha \in I} O_\alpha$, where for each $\alpha \in I$, O_α is open. Then $a \in O_\beta$ for some $\beta \in I$ and O_β is a neighborhood of a. Since $O_\beta \subset O$, by $N3$, O is a neighborhood of a. Therefore O is a neighborhood of each of its points.

Definition 6.5 A subset F of a metric space is said to be *closed* if its complement, $C(F)$, is open.

In the real number system, a closed interval $[a, b]$ is a closed set, for its complement is the union of the two open sets O_1 and O_2, where O_1 is the set of real numbers x such that $x < a$ and O_2 is the set of real numbers x such that $x > b$. A common mistake is the assumption that a set cannot be both open and closed. In any metric space (X, d), the two sets $\emptyset$ and X are open, and therefore their complements X and $\emptyset$ are closed. Thus, X and also $\emptyset$ are both open and both closed. Whether or not, in a given metric space, there are other subsets that are simultaneously open and closed, is a significant topological property, which we shall subsequently describe by the adjective "connected." In any event, the adjectives *open* and *closed* are not mutually exclusive. Nor, for that matter, are they all-inclusive, for we shall shortly give an example of a subset of the real number system that is neither open nor closed.

Definition 6.6 Let A be a subset of a metric space X. A point $b \in X$ is called a *limit point* of A if every neighborhood of b contains a point of A different from b.

If b is a limit point of A then each of the open balls $B\left(b; \frac{1}{n}\right)$ contains a point $a_n \in A$ and $\lim_n a_n = b$. Thus a limit point of a set is the limit of a convergent sequence of points of A. The converse is false, for the point b may be a point of A while for some δ, $B(b; \delta)$ contains no point of A other than b. Thus b is not a limit point of A although the sequence $b, b, \ldots$ converges to b. In this latter case b is called an *isolated point* of A.

THEOREM 6.7 In a metric space X, a set $F \subset X$ is closed if and only if F contains all its limit points.

Proof. Let F' denote the set of limit points of F. First suppose F is closed and consequently $C(F)$ is open. Choose a point $b \notin F$. Since $C(F)$ is open there is a $\delta > 0$ such that $B(b; \delta) \subset C(F)$ or $B(b; \delta) \cap F = \varnothing$. Hence $b \notin F'$ and $F' \subset F$.

Conversely, suppose $F' \subset F$, or equivalently, $C(F) \subset C(F')$. If $b \in C(F)$, then $b \notin F'$. It follows that for some $\delta > 0$, $B(b; \delta) \cap F = \varnothing$, or $B(b; \delta) \subset C(F)$. Hence $C(F)$ is open and F is closed.

THEOREM 6.8 In a metric space (X, d), a set $F \subset X$ is closed if and only if for each sequence $a_1, a_2, \ldots$ of points of F that converges to a point $a \in X$ we have $a \in F$.

Proof. First, let F be closed. Suppose $\lim_n a_n = a$ and $a_n \in F$ for $n = 1, 2, \ldots$. If the set of points $\{a_1, a_2, \ldots\}$ is infinite then every neighborhood of a contains infinitely many points of F, a is a limit point of F, and so by Theorem 6.7, $a \in F$. If this set of points is finite, then for some integer N, $a_n = a_m$ whenever $n, m > N$. Since $\lim_n a_n = a$, $d(a_n, a) = 0$ for $n > N$ or $a_n = a$, whence $a \in F$. Conversely, suppose that F is a set such that for each sequence with $\lim_n a_n = a$ and $a_n \in F$ for all n, we have $a \in F$. If b is a limit point of F then b is the limit of a convergent sequence of points of F and $b \in F$. Thus by Theorem 6.7 F is closed.

Finally, we may characterize closed sets in terms of distance from a point to a set.

THEOREM 6.9 A subset F of a metric space (X, d) is closed if and only if for each point $x \in X$, $d(x, F) = 0$ implies $x \in F$.

Proof. First, suppose F is closed. Let $x \in X$ be such that $d(x, F) = 0$. By Corollary 5.9 there is a sequence of points of F such that $\lim_n d(x, a_n) = 0$. Thus, every neighborhood of x contains points of F. If some $a_n = x$, x is in F. Otherwise each a_n is different from x, so that x is a limit point of the sequence and hence of F. Thus, by Theorem 6.7, $x \in F$. Conversely, suppose that F is such that $d(x, F) = 0$ implies $x \in F$. If x is a limit point of F then $d(x, F) = 0$. Thus in this case F contains all its limit points and is closed.

Continuity may be characterized by means of closed sets.

Theorem 6.10 Let (X, d), (Y, d') be metric spaces. A function $f{:}X \to Y$ is continuous if and only if for each closed subset A of Y, the set $f^{-1}(A)$ is a closed subset of X.

Proof. For $A \subset Y$, we have $C(f^{-1}(A)) = f^{-1}(C(A))$. But f is continuous if and only if the inverse image of each open set is an open set, and this is true if and only if the inverse image of each closed set is a closed set.

As a final result in this section we record the following facts about closed sets.

Theorem 6.11 Let (X, d) be a metric space.

*C*1. X is closed.

*C*2. $\emptyset$ is closed.

*C*3. The union of a finite collection of closed sets is closed.

*C*4. The intersection of a family of closed sets is closed.

Proof. *C*1 and *C*2 have already been discussed. *C*3 and *C*4 follow from the application of DeMorgan's formulas to the corresponding properties *O*3 and *O*4 of open sets.

The union of closed sets need not, in general, be a closed set, as may be seen by the following example. For each positive integer n let F_n be the closed interval $\left[\frac{1}{n}, 1\right]$. Then $\bigcup_{n=1}^{\infty} F_n = (0, 1]$, where $(0, 1]$ is the set of real numbers x such that $0 < x \leqq 1$. The set $(0, 1]$ is not closed, for 0 is a limit point of the set but is not in the set.

EXERCISES

1. Let (X_i, d_i), $i = 1, 2, \ldots, n$ be metric spaces. Let $X = \prod_{i=1}^{n} X_i$ and let (X, d) be the metric space defined in the standard manner by Theorem 2.3. For $i = 1, 2, \ldots, n$, let O_i be an open subset of X_i. Prove that the subset $O_1 \times O_2 \times \ldots \times O_n$ of X is open and that each open subset of X is a union of sets of this form. [A collection of open sets of a metric space is called a *basis for the open sets* if each open set is a union of sets in this collection. For example, the open balls in a metric space form a basis for the open sets.]
2. Let X be a set and d the distance function on X defined by $d(x, x) = 0$, $d(x, y) = 1$ for $x \neq y$. Prove that each subset of (X, d) is open.
3. Let (X, d_1), (Y, d_2) be metric spaces. Let $f: X \to Y$ be continuous. Define a distance function d on $X \times Y$ in the standard manner. Prove that the graph Γ_f of f is a closed subset of $(X \times Y, d)$.
4. Let $f: R \to R$ be defined by

$$f(x) = \frac{1}{x}, \; x > 0,$$

$$f(x) = 0, \; x \leqq 0.$$

Prove that the graph Γ_f is a closed subset of (R^2, d), but that f is not continuous.
5. Let A be a closed, non-empty subset of the real numbers that has a lower bound. Prove that A contains its greatest lower bound.
6. Let A be a subset of a metric space. Let A' be the set of limit points of A and A^i the set of isolated points of A. Prove that $A' \cap A^i = \emptyset$ and $A \subset A' \cup A^i$. The set $\bar{A} = A' \cup A^i$ is called the *closure* of A. Prove that $x \in \bar{A}$ if and only if there is a sequence of points of A

which converges to x. Prove that if F is a closed set such that $A \subset F$ then $\overline{A} \subset F$. Prove that $\overline{A}$ is the intersection of all such closed sets F and hence is closed.

7 SUBSPACES AND EQUIVALENCE OF METRIC SPACES

Let (X, d) be a metric space. Given a non-empty subset Y of X we may convert Y into a metric space by restricting the distance function d to $Y \times Y$. In this manner each non-empty subset Y of X gives rise to a new metric space $(Y, d \mid Y \times Y)$. On the other hand, we may be given two metric spaces (X, d) and (Y, d'). If $Y \subset X$, it makes sense to ask whether or not d' is the restriction of d.

DEFINITION 7.1 Let (X, d) and (Y, d') be metric spaces. We say that (Y, d') is a *subspace* of (X, d) if:

1. $Y \subset X$;
2. $d' = d \mid Y \times Y$.

Let $Y \subset X$ and $i: Y \to X$ be an inclusion mapping. Denote by $i \times i: Y \times Y \to X \times X$ the inclusion mapping defined by $(i \times i)(y_1, y_2) = (y_1, y_2)$. Then (Y, d') is a subspace of (X, d) if the diagram

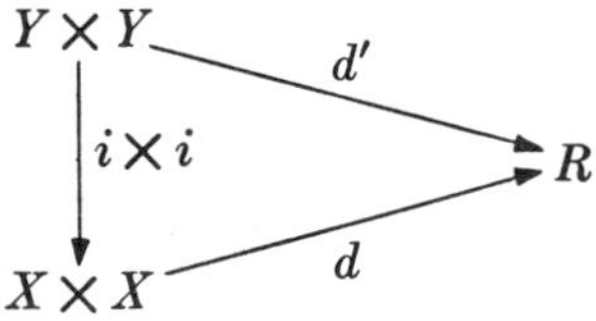

is commutative. There are as many subspaces of a metric space

(X, d) as there are non-empty subsets of X.

EXAMPLE 1 Let Q be the set of rational numbers. Define $d_Q: Q \times Q \to R$ by $d_Q(a, b) = |a - b|$. Then (Q, d_Q) is a subspace of (R, d).

EXAMPLE 2 Let I^n (the unit n-cube) be the set of all n-tuples $(x_1, x_2, \ldots, x_n)$ of real numbers such that $0 \leq x_i \leq 1$, for $i = 1, 2, \ldots, n$. Define $d_c: I^n \times I^n \to R$ by $d_c((x_1, x_2, \ldots, x_n), (y_1, y_2, \ldots, y_n)) = \underset{1 \leq i \leq n}{\text{maximum}} \{|x_i - y_i|\}$. Then (I^n, d_c) is a subspace of (R^n, d).

EXAMPLE 3 Let S^n (the n-sphere) be the set of all $(n + 1)$-tuples $(x_1, x_2, \ldots, x_{n+1})$ of real numbers such that $x_1^2 + x_2^2 + \ldots + x_{n+1}^2 = 1$. Define $d_S: S^n \times S^n \to R$ by

$$d_S((x_1, x_2, \ldots, x_{n+1}), (y_1, y_2, \ldots, y_{n+1})) = \sqrt{\sum_{i=1}^{n+1} (x_i - y_i)^2}.$$

Then (S^n, d_S) is a subspace of the Euclidean space (R^{n+1}, d').

EXAMPLE 4 Let A be the set of all $(n + 1)$-tuples $(x_1, x_2, \ldots, x_{n+1})$ of real numbers such that $x_{n+1} = 0$. Define $d_A: A \times A \to R$ by

$$d_A((x_1, x_2, \ldots, x_n, 0), (y_1, y_2, \ldots, y_n, 0)) = \underset{1 \leq i \leq n}{\text{maximum}} \{|x_i - y_i|\}.$$

Then (A, d_A) is a subspace of (R^{n+1}, d).

THEOREM 7.2 Let (Y, d') be a subspace of (X, d). Then the inclusion mapping $i: Y \to X$ is continuous.

Proof. Given $a \in Y$ and $\varepsilon > 0$, choose $\delta = \varepsilon$. If $d'(a, y) < \delta$, then $d(i(a), i(y)) = d(a, y) = d'(a, y) < \delta = \varepsilon$.

The metric space (A, d_A) of Example 4 is in most respects a copy of the metric space (R^n, d). The only distinction between (R^n, d) and (A, d_A) is that a point of R^n is an n-tuple of real numbers, whereas a point of A is an $(n + 1)$-tuple of real numbers of which the last one is zero. The relationship between the metric spaces (R^n, d) and (A, d_A) is an example of the relationship called "metric equivalence" or "isometry."

Definition 7.3 Two metric spaces (A, d_A) and (B, d_B) are said to be *metrically equivalent* or *isometric* if there are inverse functions $f:A \to B$ and $g:B \to A$ such that, for each $x, y \in A$, $d_B(f(x), f(y)) = d_A(x, y)$, and for each $u, v \in B$, $d_A(g(u), g(v)) = d_B(u, v)$. In this event we shall say that the *metric equivalence* or *isometry is defined by f and g.*

Theorem 7.4 A necessary and sufficient condition that two metric spaces (A, d_A) and (B, d_B) be metrically equivalent is that there exist a function $f:A \to B$ such that:

1. f is one-one;
2. f is onto;
3. for each $x, y \in A$, $d_B(f(x), f(y)) = d_A(x, y)$.

Proof. The stated conditions are necessary, for if (A, d_A) and (B, d_B) are metrically equivalent, there are inverse functions $f:A \to B$ and $g:B \to A$, and therefore f is one-one and onto. Conversely, suppose a function $f:A \to B$ with the stated properties exists. Then f is invertible and the function $g:B \to A$ such that f and g are inverse functions is determined by setting $g(b) = a$ if $f(a) = b$. For $u, v \in B$, let $x = g(u)$, $y = g(v)$. Then

$$d_A(g(u), g(v)) = d_A(x, y) = d_B(f(x), f(y)) = d_B(u, v).$$

Given metric spaces (A, d_A) and (B, d_B) and functions $f:A \to B$ and $g:B \to A$, let us denote by $f \times f:A \times A \to B \times B$ the function defined by setting $(f \times f)(x, y) = (f(x), f(y))$ for $x, y \in A$ and, similarly, let $g \times g:B \times B \to A \times A$ be defined by setting $(g \times g)(u, v) = (g(u), g(v))$ for $u, v \in B$. The statement that $d_B(f(x), f(y)) = d_A(x, y)$ for $x, y \in A$ is equivalent to the statement that the diagram

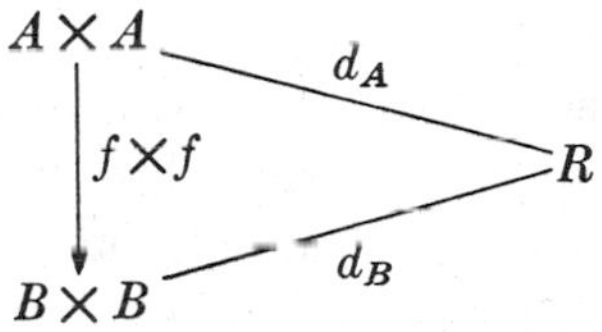

is commutative (onc may also describe this relation by saying that the function $f: A \to B$ is "distance preserving"). In terms of diagrams, the statement that (A, d_A) and (B, d_B) are metrically equivalent is the statement that there exist functions $f: A \to B$, $g: B \to A$ such that thc four diagrams

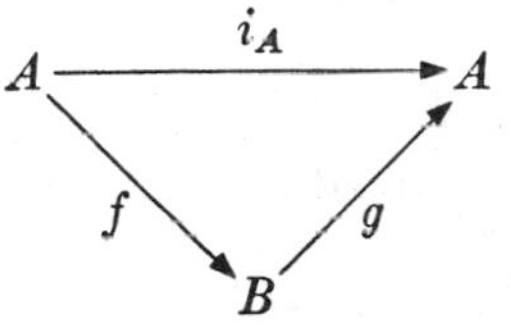

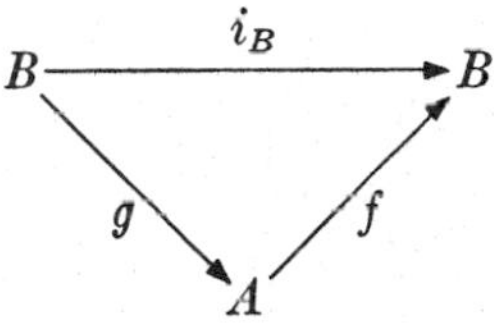

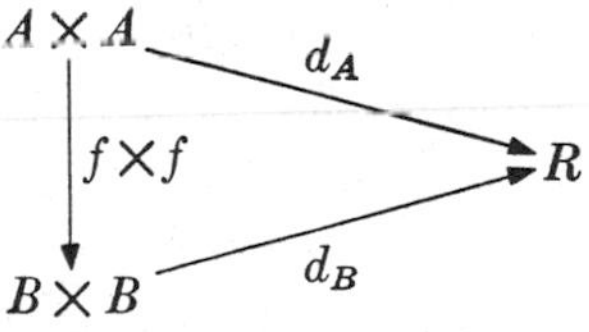

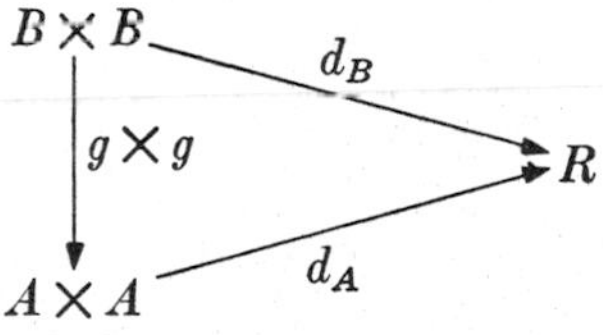

are commutative (where $i_A: A \to A$ and $i_B: B \to B$ are identity mappings). The first two diagrams express the fact that f and g are inverse functions and the last two diagrams express the fact that f and g "preserve distances." Since the distance between x and y in A is the same as the distance between $f(x)$ and $f(y)$ in B, f is continuous. Similarly, g is continuous. Thus:

LEMMA 7.5 Let a metric equivalence between (A, d_A) and (B, d_B) be defined by inverse functions $f:A \to B$ and $g:B \to A$. Then both f and g are continuous.

From the point of view of considerations that relate only to the concept of continuity, the relationship of metric equivalence is too narrow. We are led to define a broader concept of equivalence in which we drop the requirement of "preservation of distance"; that is, the commutativity of the last pair of diagrams, and merely require that the first two diagrams be commutative and the functions in these diagrams be continuous.

DEFINITION 7.6 Two metric spaces (A, d_A) and (B, d_B) are said to be *topologically equivalent* if there are inverse functions $f:A \to B$ and $g:B \to A$ such that f and g are continuous. In this event we say that the *topological equivalence is defined by f and g.*

As a corollary to Lemma 7.5 we obtain:

COROLLARY 7.7 Two metric spaces that are metrically equivalent are topologically equivalent.

The converse of this corollary is false; that is, there are metric spaces that are topologically equivalent, but are not metrically equivalent. For example, a circle of radius 1 is topologically equivalent to a circle of radius 2 (considered as subspaces of (R^2, d)), but the two are not metrically equivalent.

The following two results furnish a sufficient condition for the topological equivalence of two metric spaces with the same underlying sets.

LEMMA 7.8 Let (X, d_1) and (X, d_2) be two metric spaces. If there exists a number $K > 0$ such that for each $x, y \in X$, $d_2(x, y) \leqq Kd_1(x, y)$, then the identity mapping

$$i:(X, d_1) \to (X, d_2)$$

is continuous.

Proof. Given $\varepsilon > 0$ and $a \in X$, set $\delta = \varepsilon/K$. If $d_1(x, a) < \delta$ then $d_2(i(x), i(a)) = d_2(x, a) \leqq K \cdot d_1(x, a) < K\delta = \varepsilon$.

COROLLARY 7.9 Let (X, d) and (X, d') be two metric spaces with the same underlying set. If there exist positive numbers K and K' such that for each $x, y \in X$,

$$d'(x, y) \leqq K \cdot d(x, y),$$

$$d(x, y) \leqq K' \cdot d'(x, y),$$

then the identity mappings define a topological equivalence between (X, d) and (X, d').

We have discussed the two metric spaces (R^n, d) and (R^n, d'), where the distance function d is determined by the maximum distance between coordinates, and the distance function d' is what is called the Euclidean distance function. For each pair of points $x, y \in R^n$, the inequality $d(x, y) \leqq d'(x, y) \leqq \sqrt{n}\, d(x, y)$ holds. It therefore follows from Corollary 7.9 that the metric spaces (R^n, d) and (R^n, d') are topologically equivalent.

THEOREM 7.10 Let (X, d) and (Y, d') be two metric spaces. Let $f: X \to Y$ and $g: Y \to X$ be inverse functions. Then the following four statements are equivalent:

1. f and g are continuous;
2. A subset O of X is open if and only if $f(O)$ is an open subset of Y;
3. A subset F of X is closed if and only if $f(F)$ is a closed subset of Y;
4. For each $a \in X$ and subset N of X, N is a neighborhood of a if and only if $f(N)$ is a neighborhood of $f(a)$.

Proof. $1 \Rightarrow 2$. Let O be an open subset of X. Then $f(O) = g^{-1}(O)$ is open since g is continuous. Conversely, if $f(O)$ is an open subset of Y, then $f^{-1}(f(O)) = O$ is open since f is continuous.

$2 \Rightarrow 4$. For each $a \in X$ and $N \subset X$, N is a neighborhood of a if and only if N contains an open set O

containing a if and only if $f(N)$ contains an open set $O' = f(O)$ containing $f(a)$ if and only if $f(N)$ is a neighborhood of $f(a)$.

$4 \Rightarrow 1$. Let $a \in X$ and let U be a neighborhood of $f(a)$. Then $f^{-1}(U)$ is a neighborhood of a, for $U = f(f^{-1}(U))$ is a neighborhood of $f(a)$. Thus f is continuous. Similarly, let $b \in Y$ and let V be a neighborhood of $g(b)$. Then $g^{-1}(V) = f(V)$ is a neighborhood of $f(g(b)) = b$, and g is continuous.

Thus, statements 1, 2, and 4 are equivalent. We leave it to the reader to verify that statements 2 and 3 are equivalent.

Statement 1 in Theorem 7.10 is, of course, the statement that the metric spaces (X, d) and (Y, d') are topologically equivalent. Consequently, Theorem 7.10 asserts that two metric spaces are topologically equivalent if and only if there exist inverse functions that establish either a one-one correspondence between the open sets of the two spaces, a one-one correspondence between the closed sets of the two spaces, or a one-one correspondence between the complete systems of neighborhoods of the two spaces.

Both metrically equivalent and topologically equivalent are equivalence relations defined on a collection of metric spaces. By Corollary 7.7, each equivalence class of metrically equivalent metric spaces is contained in an equivalence class of topologically equivalent metric spaces. Distinguishing which topologically equivalent equivalence class a metric space belongs to is a coarser, but consequently more fundamental, distinction. By Theorem 7.10, this is determined by the collection of open sets, or the "topology" of the space.

EXERCISES

1. For each pair of points $a, b \in R^n$, prove that there is a topological equivalence between (R^n, d) and itself defined by inverse functions $f: R^n \to R^n$ and $g: R^n \to R^n$ such that $f(a) = b$. [*Hint:* If $a = (a_1, a_2, \ldots, a_n)$, $b = (b_1, b_2, \ldots, b_n)$, define f by setting

$f(x_1, x_2, \ldots, x_n) = (x_1 + b_1 - a_1, x_2 + b_2 - a_2, \ldots, x_n + b_n - a_n)$.]

2. Prove that the open interval $(-\pi/2, \pi/2)$, considered as a subspace of the real number system, is topologically equivalent to the real number system. Prove that any two open intervals, considered as subspaces of the real number system, are topologically equivalent. Prove that any open interval, considered as a subspace of the real number system, is topologically equivalent to the real number system.
3. For $i = 1, 2, \ldots, n$, let the metric space (X_i, d_i) be topologically equivalent to the metric space (Y_i, d_i'). Prove that if

$$X = \prod_{i=1}^{n} X_i \quad \text{and} \quad Y = \prod_{i=1}^{n} Y_i$$

are converted into metric spaces in the standard manner, then these two metric spaces are topologically equivalent.
4. The open n-cube is the set of all points $x = (x_1, x_2, \ldots, x_n) \in R^n$ such that $0 < x_i < 1$ for $i = 1, 2, \ldots, n$. Prove that the open n-cube, considered as a subspace of (R^n, d), is topologically equivalent to (R^n, d). [*Hint:* Use the results of Problems 2 and 3.]
5. Let XRY mean that the metric space X is isometric to the metric space Y. Prove that: (i) XRX; (ii) if XRY then YRX; and (iii) if XRY and YRZ then XRZ. Do the same if XRY means that the metric space X is topologically equivalent to the metric space Y.
6. Let (Y, d') be a subspace of the metric space (X, d). Prove that a subset $O' \subset Y$ is an open subset of (Y, d') if and only if there is an open subset O of (X, d) such that $O' = Y \cap O$. Prove that a subset $F' \subset Y$ is a closed subset of (Y, d') if and only if there is a closed subset F of (X, d) such that $F' = Y \cap F$. For a point $a \in Y$, prove that a subset $N' \subset Y$ is a neighborhood of a if and only if there is a neighborhood N of a in (X, d) such that $N' = Y \cap N$.
7. Let (Y, d') be a subspace of (X, d). Let $a_1, a_2, \ldots$ be a sequence of points of Y and let $a \in Y$. Prove that if $\lim_n a_n = a$ in (Y, d'), then $\lim_n a_n = a$ in (X, d). [The converse is false unless one assumes that all the points mentioned lie in Y; see the next problem.]
8. Consider the subspace (Q, d_Q) (the rational numbers) of (R, d). Let $a_1, a_2, \ldots$ be a sequence of rational numbers such that $\lim_n a_n = \sqrt{2}$. Prove that, given $\varepsilon > 0$, there is a positive integer N such that for $n, m > N$, $|a_n - a_m| < \varepsilon$. Does the sequence $a_1, a_2, \ldots$ converge when considered to be a sequence of points of (Q, d_Q)?

8 AN INFINITE DIMENSIONAL EUCLIDEAN SPACE

In this section we shall define a metric space H, sometimes called Hilbert space, which contains as subspaces isometric copies of the various Euclidean spaces (R^n, d'). A point u of H is a sequence $u_1, u_2, \ldots$ of real numbers such that the series $\sum_{i=1}^{\infty} u_i^2$ is convergent. Let $u = (u_1, u_2, \ldots)$ and $v = (v_1, v_2, \ldots)$ be in H. Our intention is to define a metric on H by setting

$$d(u, v) = \left[\sum_{i=1}^{\infty} (u_i - v_i)^2\right]^{1/2}.$$

In order to do this we must first know that the series in brackets converges. To accomplish this we shall make use of the following result, which is frequently referred to as Schwarz's lemma or Cauchy's inequality.

LEMMA 8.1 Let $(u_1, u_2, \ldots, u_n)$, $(v_1, v_2, \ldots, v_n)$ be n-tuples of real numbers, then

$$\sum_{i=1}^{n} u_i v_i \leqq \left[\sum_{i=1}^{n} u_i^2\right]^{1/2} \left[\sum_{i=1}^{n} v_i^2\right]^{1/2}.$$

Proof. It suffices to prove that

$$\left(\sum_{i=1}^{n} u_i v_i\right)^2 \leqq \left(\sum_{i=1}^{n} u_i^2\right)\left(\sum_{i=1}^{n} v_i^2\right).$$

To this end, we consider, for an arbitrary real number λ, the expression $\sum_{i=1}^{n} (u_i + \lambda v_i)^2$. We have,

$$0 \leqq \sum_{i=1}^{n} (u_i + \lambda v_i)^2 = \sum_{i=1}^{n} u_i^2 + 2\lambda \sum_{i=1}^{n} u_i v_i + \lambda^2 \sum_{i=1}^{n} v_i^2.$$

Therefore, the quadratic equation in λ,

$$0 = \sum_{i=1}^{n} u_i^2 + 2\lambda \sum_{i=1}^{n} u_i v_i + \lambda^2 \sum_{i=1}^{n} v_i^2,$$

can have at most one real solution. Consequently,

$$\left(\sum_{i=1}^{n} u_i v_i\right)^2 - \left(\sum_{i=1}^{n} u_i^2\right)\left(\sum_{i=1}^{n} v_i^2\right) \leqq 0,$$

or

$$\left(\sum_{i=1}^{n} u_i v_i\right)^2 \leqq \left(\sum_{i=1}^{n} u_i^2\right)\left(\sum_{i=1}^{n} v_i^2\right).$$

COROLLARY 8.2 Let $u = (u_1, u_2, \ldots)$, $v = (v_1, v_2, \ldots)$ be in H with $U = \sum_{i=1}^{\infty} u_i^2$, $V = \sum_{i=1}^{\infty} v_i^2$. Then the series $\sum_{i=1}^{\infty} u_i v_i$ is absolutely convergent and $\sum_{i=1}^{\infty} |u_i v_i| \leqq U^{1/2}V^{1/2}$.

Proof. For each positive integer n

$$\sum_{i=1}^{n} |u_i v_i| = \sum_{i=1}^{n} |u_i|\,|v_i| \leqq \left[\sum_{i=1}^{n} |u_i|^2\right]^{1/2}\left[\sum_{i=1}^{n} |v_i|^2\right]^{1/2} \leqq U^{1/2}V^{1/2}.$$

Thus the partial sums of this series of positive terms are bounded and the series converges to a limit not greater than $U^{1/2}V^{1/2}$.

Furthermore, if α and β are real numbers and we set $\alpha u + \beta v = (\alpha u_1 + \beta v_1, \alpha u_2 + \beta v_2, \ldots)$ then $\alpha u + \beta v$ is also in H for $\sum_{i=1}^{\infty} (\alpha u_i + \beta v_i)^2$ is the sum of three absolutely convergent series. In particular $u + v \in H$ and

$$\sum_{i=1}^{\infty} (u_i + v_i)^2 = \sum_{i=1}^{\infty} |u_i^2 + 2u_i v_i + v_i^2| \leqq \sum_{i=1}^{\infty} u_i^2 + 2\sum_{i=1}^{\infty} |u_i v_i| + \sum_{i=1}^{\infty} v_i^2$$
$$\leqq U + 2U^{1/2}V^{1/2} + V = (U^{1/2} + V^{1/2})^2.$$

Taking square roots we obtain

COROLLARY 8.3 $\left[\sum_{i=1}^{\infty} (u_i + v_i)^2\right]^{1/2} \leqq U^{1/2} + V^{1/2}.$

THEOREM 8.4 (H, d) is a metric space, where d is defined by $d(u, v) = \left[\sum_{i=1}^{\infty} (u_i - v_i)^2\right]^{1/2}$.

Proof. It is readily apparent that d satisfies all the properties of a distance function with the exception of the property that $d(a, b) \leqq d(a, c) + d(c, b)$ for a, b,

$c \in H$. Let $a = (a_1, a_2, \ldots)$, $b = (b_1, b_2, \ldots)$, $c = (c_1, c_2, \ldots)$. Set $u = a - c$, $v = c - b$ so that $u_i = a_i - c_i$, $v_i = c_i - b_i$. Then $u_i + v_i = a_i - b_i$ and Corollary 8.3 yields the desired inequality.

Let E^n be the collection of points $u = (u_1, u_2, \ldots) \in H$ such that $u_j = 0$ for $j > n$. To each point $a = (a_1, a_2, \ldots, a_n) \in R^n$ we can associate the point $h(a) = (a_1, a_2, \ldots, a_n, 0, 0, \ldots) \in E^n$. Clearly h is a one-one mapping of R^n onto the subspace E^n of H. Using $d'(a, b) = \left[\sum_{i=1}^{n} (a_i - b_i)^2\right]^{1/2}$ in R^n, $d'(a, b) = d(h(a), h(b))$. Since E^n is a metric space, (R^n, d') is a metric space and h is an isometry of (R^n, d') with $(E^n, d|E^n)$.

EXERCISES

1. Let V be a vector space with the real numbers R as scalars. A function $A: V \times V \to R$ is called a *bilinear form* if $A(\alpha a + \beta b, c) = \alpha A(a, c) + \beta A(b, c)$ and $A(a, \beta b + \gamma c) = \beta A(a, b) + \gamma A(a, c)$ for scalars α, β, and $\gamma \in R$ and vectors a, b, and $c \in V$. A bilinear form is called *positive definite* if $A(x, x) > 0$, unless x is the zero vector. Define a vector space structure on Hilbert space H and show that for $u = (u_1, u_2, \ldots)$ and $v = (v_1, v_2, \ldots) \in H$, $A(u, v) = \sum_{i=1}^{\infty} u_i v_i$ yields a positive definite bilinear form.
2. Let V be a vector space with the real numbers R as scalars. A *norm* on V is a function $N: V \to R$ such that (i) $N(v) \geqq 0$ for all $v \in V$; (ii) $N(v) = 0$ if and only if $v = 0$; (iii) $N(u + v) \leqq N(u) + N(v)$ for all $u, v \in V$; (iv) $N(\alpha v) = |\alpha| N(v)$ for all $\alpha \in R$, $v \in V$. Prove that if A is a positive definite bilinear form on V, then $N(v) = (A(v, v))^{1/2}$ defines a norm on V.
3. Let N be a norm on a vector space V as defined in the previous problem. Set $d(u, v) = N(u - v)$ for $u, v \in V$. Prove that (V, d) is a metric space. Prove that the following functions are continuous: (i) $a: V \times V \to V$ defined by $a(u, v) = u + v$; (ii) $b: V \to V$ defined by $b(v) = -v$; (iii) $c: R \times V \to V$ defined by $c(\alpha, v) = \alpha v$.

For further reading, Kaplansky, *Set Theory and Metric Spaces*, Kolmogorov and Fomin, *Elements of the Theory of Functions and Functional Analysis*, and Simmons, *Introduction to Topology and Modern Analysis* all have excellent chapters on metric spaces.

CHAPTER 3

Topological Spaces

1 INTRODUCTION

In the context of metric spaces, the various topological concepts such as continuity, neighborhood, and so on, may be characterized by means of open sets. Discarding the distance function and retaining the open sets of a metric space gives rise to a new mathematical object, called a *topological space*. The topological concepts that have been studied in Chapter 2 must be reintroduced in the context of topological spaces. The procedure for formulating the appropriate definitions of these terms in a topological space is to find, in a metric space, the characterization of the term by means of open sets, using in most cases what is a theorem in a metric space as a definition in a topological space. There are other ways of introducing topological spaces. For example, if, upon discarding the distance function of a metric space, we were to retain the systems of neighborhoods of the points of the metric space, we obtain what we shall call a neighborhood space. We shall indicate the equivalence between the concept of a neighborhood space and the concept of a topological space.

Certain new topological concepts are also introduced; namely, the closure, interior, and boundary of a set (these concepts could have been introduced in metric spaces). In many respects the elementary material in this chapter is a repetition of material from Chapter 2, but in a different context. The concept of a topological space is one of the most fruitful concepts of modern mathematics. It is the proper setting for discussions based on considerations of continuity.

2 TOPOLOGICAL SPACES

DEFINITION 2.1 Let X be a non-empty set and $\mathfrak{I}$ a collection of subsets of X such that:

$O1$. $X \in \mathfrak{I}$.
$O2$. $\emptyset \in \mathfrak{I}$.
$O3$. If $O_1, O_2, \ldots, O_n \in \mathfrak{I}$, then

$$O_1 \cap O_2 \cap \ldots \cap O_n \in \mathfrak{I}.$$

$O4$. If for each $\alpha \in I$, $O_\alpha \in \mathfrak{I}$, then $\bigcup_{\alpha \in I} O_\alpha \in \mathfrak{I}$.

The pair of objects $(X, \mathfrak{I})$ is called a *topological space*. The set X is called the *underlying set*, the collection $\mathfrak{I}$ is called the *topology* on the set X, and the members of $\mathfrak{I}$ are called *open* sets.

By virtue of Theorem 6.4, Chapter 2, if $\mathfrak{I}$ is the collection of open sets of a metric space (X, d), then $(X, \mathfrak{I})$ is a topological space, called the *topological space associated with the metric space* (X, d), and the metric space (X, d) is said to *give rise to the topological space* $(X, \mathfrak{I})$. We are therefore in a position to give many examples of topological spaces; namely, for each metric space its associated topological space. On the other hand, any set X and collection $\mathfrak{I}$ of subsets satisfying $O1$, $O2$, $O3$, $O4$ is an example of a topological space, and we shall see that not every such example arises from a metric space.

EXAMPLES

1. The *real line*, that is, the topological space that arises from the metric space consisting of the real number system and the distance function $d(a, b) = |a - b|$.
2. The topological space that arises from the metric space (R^n, d). We shall call this topological space *Euclidean n-space with the usual topology*.
3. Let X be an arbitrary set. Let $\mathfrak{I} = \{\emptyset, X\}$. Then $(X, \mathfrak{I})$ is a topological space.
4. Let X be a set containing precisely two distinct elements a and b. Let $\mathfrak{I}_1 = \{\emptyset, X\}$, $\mathfrak{I}_2 = \{\emptyset, \{a\}, X\}$, $\mathfrak{I}_3 = \{\emptyset, \{b\}, X\}$, $\mathfrak{I}_4 = \{\emptyset, \{a\}, \{b\}, X\}$. Then $(X, \mathfrak{I}_i)$, $i = 1, 2, 3, 4$, are four distinct topological spaces with the same underlying set.
5. Let X be an arbitrary set. Let $\mathfrak{I}$ be the collection of all subsets of X, i.e., $\mathfrak{I} = 2^X$. Then $(X, \mathfrak{I})$ is a topological space. Of all the various topologies that one may place on a set X, this one contains the largest number of elements and is called the *discrete* topology.
6. Let X be an arbitrary set. Let $\mathfrak{I}$ be the collection of all subsets of X whose complements are either finite or all of X. Then $(X, \mathfrak{I})$ is a topological space.
7. Let Z be the set of positive integers. For each positive integer n, let $O_n = \{n, n + 1, n + 2, \ldots\}$. Let $\mathfrak{I} = \{\emptyset, O_1, O_2 \ldots, O_n, \ldots,\}$. Then $(Z, \mathfrak{I})$ is a topological space.

To verify that $(X, \mathfrak{I})$ is a topological space, one verifies that the specified collection of subsets, $\mathfrak{I}$, is a topology; that is, that $\mathfrak{I}$ satisfies conditions $O1$, $O2$, $O3$, $O4$. For example, let X and $\mathfrak{I}$ be as in Example 6. Then $X \in \mathfrak{I}$, for its complement $\emptyset = C(X)$ is certainly finite. Also $\emptyset \in \mathfrak{I}$, since $C(\emptyset) = X$. Thus, $\mathfrak{I}$ satisfies conditions $O1$ and $O2$. Next, let $O_1, O_2, \ldots, O_n$ be subsets of X, each of whose complements is finite or all of X. To show that $O_1 \cap O_2 \cap \ldots \cap O_n \in \mathfrak{I}$ we must show that $C(O_1 \cap O_2 \cap \ldots \cap O_n)$ is either finite or all of X. But $C(O_1 \cap O_2 \cap \ldots \cap O_n) = C(O_1) \cup C(O_2) \cup \ldots \cup C(O_n)$. Either this set is a union of finite sets and hence finite, or for some i, $C(O_i) = X$ and the union is all of X. Finally, for each

$\alpha \in I$, let $O_\alpha \in \mathfrak{J}$, so that $C(O_\alpha)$ is either finite or X. Then $C(\bigcup_{\alpha \in I} O_\alpha) = \bigcap_{\alpha \in I} C(O_\alpha)$. Either each of the sets, $C(O_\alpha) = X$, in which case the intersection is all of X, or at least one of them is finite, in which case the intersection is a subset of a finite set and hence finite. Thus $(X, \mathfrak{J})$ is a topological space. The reader should verify that the remaining examples do, in fact, constitute examples of topological spaces.

The relationship between the totality of metric spaces and the totality of topological spaces is indicated in Figure 8. We shall see that two distinct metric spaces (X, d) and (X, d') may give rise to the same topological space $(X, \mathfrak{J})$. Also there are topological spaces $(Y, \mathfrak{J}')$, such as Example 7 above, which could not have arisen from a metric space. The subcollection of topological spaces that arise from metric spaces is called the collection of *metrizable* topological spaces. In passing from a metric space

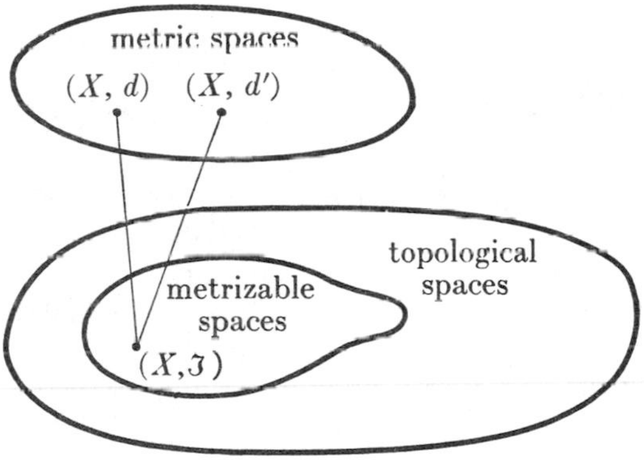

Figure 8

to its associated topological space, we may say that the "open" sets have been "preserved."

Definition 2.2 Given a topological space $(X, \mathfrak{J})$, a subset N of X is called a *neighborhood* of a point $a \in X$ if N contains an open set that contains a.

This definition has been formulated so that a subset N of a metric space (X, d) is a neighborhood of a point $a \in X$ if and only if N is a neighborhood of a in the associated topological space. Thus, in passing from a metric space to a topological space, neighborhoods have also been "preserved."

COROLLARY 2.3 Let $(X, \mathfrak{I})$ be a topological space. A subset O of X is open if and only if O is a neighborhood of each of its points.

Proof. First, suppose that O is open. Then, for each $x \in O$, O contains an open set containing x; namely, O itself. Conversely, suppose O is a neighborhood of each of its points. Then for each $x \in O$, there is an open set O_x such that $x \in O_x \subset O$. Consequently, $O = \bigcup_{x \in O} O_x$ is a union of open sets and hence open.

DEFINITION 2.4 Given a topological space $(X, \mathfrak{I})$, a subset F of X is called a *closed* set if the complement, $C(F)$, is an open set.

EXERCISES

1. Let $(X, \mathfrak{I})$ be a topological space that is metrizable. Prove that for each pair a, b of distinct points of X, there are open sets O_a and O_b containing a and b respectively, such that $O_a \cap O_b = \emptyset$. Prove that the topological space of Example 7 is not metrizable.
2. Prove that for each set X, the topological space $(X, 2^X)$ is metrizable. [*Hint*: See Exercise 2, Chapter 2, Section 6.]
3. Let (R^n, d) and (R^n, d') be defined as in Chapter 2 so that for $x = (x_1, x_2, \ldots, x_n)$ and $y = (y_1, y_2, \ldots, y_n) \in R^n$,

$$d(x, y) = \underset{1 \leq i \leq n}{\text{maximum}} \{|x_i - y_i|\},$$

$$d'(x, y) = \left(\sum_{i=1}^{n} (x_i - y_i)^2\right)^{1/2}.$$

Prove that the two metric spaces (R^n, d) and (R^n, d') give rise to the same topological space.

4. Let $(X, \mathfrak{I})$ be a topological space. Prove that $\emptyset$, X are closed sets, that a finite union of closed sets is a closed set, and that an arbitrary intersection of closed sets is a closed set.

5. Let $(X, \mathfrak{I})$ be a topological space that is metrizable. Prove that each neighborhood N of a point $a \in X$ contains a neighborhood V of a such that V is a closed set.
6. Prove that in a discrete topological space, each subset is simultaneously open and closed.

3 NEIGHBORHOODS AND NEIGHBORHOOD SPACES

Theorem 4.8, Chapter 2, in which are stated certain properties of neighborhoods in a metric space, corresponds to a theorem in topological spaces.

THEOREM 3.1 Let $(X, \mathfrak{I})$ be a topological space.

$N1$. For each point $x \in X$, there is at least one neighborhood N of x.

$N2$. For each point $x \in X$ and each neighborhood N of x, $x \in N$.

$N3$. For each point $x \in X$, if N is a neighborhood of x and $N' \supset N$, then N' is a neighborhood of x.

$N4$. For each point $x \in X$ and each pair N, M of neighborhoods of x, $N \cap M$ is also a neighborhood of x.

$N5$. For each point $x \in X$ and each neighborhood N of x, there exists a neighborhood O of x such that $O \subset N$ and O is a neighborhood of each of its points.

Proof. For each point $x \in X$, X is a neighborhood of x, thus $N1$ is true. $N2$ and $N3$ follow easily from the definition of neighborhood in a topological space. To verify $N4$, let N, M be neighborhoods of x. Then there are open sets O and O' such that $N \supset O$, $M \supset O'$ and $x \in O$, $x \in O'$. Thus, $N \cap M$ contains the open set $O \cap O'$, which contains x, and, consequently, $N \cap M$ is a neighborhood of x. Finally, for a point $x \in X$, let N be a neighborhood of x. Then N contains an open set O containing x. In particular, O is a neighborhood of x and by Corollary 2.3, O is a neighborhood of each of its points.

In a topological space, as in a metric space, we lay down the definition:

Definition 3.2 For each point x in a topological space $(X, \mathfrak{J})$, the collection $\mathfrak{N}_x$ of all neighborhoods of x is called a *complete system of neighborhoods at the point x.*

One may paraphrase the properties $N1$–$N5$ of neighborhoods in terms of the complete system of neighborhoods $\mathfrak{N}_x$ at the points $x \in X$:

$N1$. For each $x \in X$, $\mathfrak{N}_x \neq \emptyset$;
$N2$. For each $x \in X$ and $N \in \mathfrak{N}_x$, $x \in N$;
$N3$. For each $x \in X$ and $N \in \mathfrak{N}_x$, if $N' \supset N$ then $N' \in \mathfrak{N}_x$;
$N4$. For each $x \in X$ and $N, M \in \mathfrak{N}_x$,

$$N \cap M \in \mathfrak{N}_x;$$

$N5$. For each $x \in X$ and $N \in \mathfrak{N}_x$, there exists an $O \in \mathfrak{N}_x$ such that $O \subset N$ and $O \in \mathfrak{N}_y$ for each $y \in O$.

The proof of Theorem 3.1 was, in most respects, similar to the proof of the corresponding theorem in metric spaces, Theorem 4.8, Chapter 2. However, it was necessary to supply a proof of Theorem 3.1 above, for in the proof of 4.8, Chapter 2, use was made of the concept of open balls, a concept which does not occur in a topological space. Though a comparison of these two theorems might lead one to believe that statements about neighborhoods that are true in a metric space are also true in a topological space, this is not always the case. Given two distinct points x and y in a metric space (X, d) there are neighborhoods N and M of x and y respectively, such that $N \cap M = \emptyset$. This statement is false in many topological spaces. For example, let $Y = \{a, b\}$, $a \neq b$, and let $\mathfrak{J} = \{\emptyset, \{a\}, Y\}$, so that $(Y, \mathfrak{J})$ is a topological space. Then the only neighborhood of b is Y. Thus, for each neighborhood N of a and each neighborhood M of b, $N \cap M = N \cap Y = N \neq \emptyset$.

Definition 3.3 A topological space $(X, \mathfrak{J})$ is called a *Hausdorff space* or is said to satisfy the *Hausdorff axiom,* if for each pair a, b of distinct points of X, there are neighborhoods N and M of a and b respectively, such that $N \cap M = \emptyset$.

Some authors use the term "separated space" instead of Hausdorff space. Many of the significant topological spaces are Hausdorff spaces. For this reason certain authors require a topological space to be a Hausdorff space and use the two terms synonymously; that is, they add to the list $O1$–$O4$ of properties of open sets in the definition of a topological space, the property, for each pair x, y of distinct points there are open sets O_x and O_y containing x and y respectively, such that $O_x \cap O_y = \emptyset$.

Suppose we have a metric space (X, d) and we discard the distance function, retaining only the neighborhoods of the points in X. Then for each point $x \in X$, we have a collection $\mathfrak{N}_x$ of subsets of X; namely the complete system of neighborhoods at x. These neighborhoods satisfy certain properties. We may select some of these properties and use them as a set of axioms for what we might naturally call a "neighborhood space."

DEFINITION 3.4 Let X be a set. For each $x \in X$, let there be given a collection $\mathfrak{N}_x$ of subsets of X (called the neighborhoods of x), satisfying the conditions $N1$–$N5$ of Theorem 3.1. This object is called a *neighborhood space.*

In a neighborhood space, the appropriate definition of *open* set is obtained from Corollary 2.3.

DEFINITION 3.5 In a neighborhood space, a subset O is said to be *open* if it is a neighborhood of each of its points.

It is important to realize that the mathematical object neighborhood space, although closely connected with the concept of a topological space, is a new object, and until we have defined the term *open set* in a neighborhood space, that term in a neighborhood space is meaningless.

LEMMA 3.6 In a neighborhood space, the empty set and the whole space are open, a finite intersection of open sets is open, and an arbitrary union of open sets is open.

Proof. [Since we are concerned with neighborhood spaces, we may use only the properties $N1$–$N5$ of neighborhoods and, of course, Definition 3.5 of open sets.] The empty set is open, for in order for it not to be open it would have to contain a point x of which it was not a neighborhood. Given a point x, there is some neighborhood N of x, so by $N3$, the whole space is a neighborhood of x. Thus, the whole space is a neighborhood of each of its points and hence open. If O and O' are open, then $O \cap O'$ is also open, for by $N4$, given $x \in O \cap O'$, O and O' are neighborhoods of x, hence so is $O \cap O'$. Thus the intersection of two open sets is a neighborhood of each of its points, and, consequently, by induction, any finite intersection of open sets is open. Finally, suppose for each $\alpha \in I$, O_α is open. If $x \in \bigcup_{\alpha \in I} O_\alpha$, then $x \in O_\beta$ for some $\beta \in I$. But O_β is a neighborhood of x and $O_\beta \subset \bigcup_{\alpha \in I} O_\alpha$, thus by $N3$, $\bigcup_{\alpha \in I} O_\alpha$ is a neighborhood of x and is therefore open.

If we start with a topological space and define neighborhoods by Definition 2.2, Theorem 3.1 tells us that the underlying set and the complete systems of neighborhoods of the points of the set yield a neighborhood space. On the other hand, if we start with a neighborhood space and define open sets by Definition 3.5, Lemma 3.6 tells us that we obtain a topological space. Suppose then, we have a topological space $(X, \mathfrak{I})$, use the neighborhoods of $(X, \mathfrak{I})$ to form a neighborhood space, and finally use the open sets in this neighborhood space to create a topological space $(X, \mathfrak{I}')$. Do we end up with our original topological space $(X, \mathfrak{I})$? The answer is yes. To prove this result we must show that $\mathfrak{I} = \mathfrak{I}'$. Now, if O is an open set in our original topological space, that is, $O \in \mathfrak{I}$, by Corollary 2.3, O is a neighborhood of each of its points, from which it follows that O is an open subset of the neighborhood space and hence $O \in \mathfrak{I}'$. Conversely, if $O \in \mathfrak{I}'$, then in the neighborhood space, O is a neighborhood of each of its points. But the neighborhoods of the neighborhood space we have created are the neighborhoods of $(X, \mathfrak{I})$, so that again by Corollary 2.3, O is open in $(X, \mathfrak{I})$ or $O \in \mathfrak{I}$. Thus $\mathfrak{I} = \mathfrak{I}'$.

Logically, it would still be possible for there to be neighborhood spaces that did not arise in this manner from topological spaces. We shall now show that there are none. To do so, we need a characterization, in a neighborhood space, of neighborhoods in terms of open sets.

LEMMA 3.7 In a neighborhood space, a subset N is a neighborhood of a point x if and only if N contains an open set containing x.

Proof. First, let N contain an open set O containing x. By Definition 3.5, O is a neighborhood of x, whence, by $N3$, N is a neighborhood of x. Conversely, if N is a neighborhood of x, then by $N5$, N contains a neighborhood O of x (and by $N2$, O contains x), such that O is a neighborhood of each of its points.

To denote a neighborhood space, let us use the symbol $(X, \mathfrak{N})$, where for each $x \in X$, $\mathfrak{N}_x$ is the collection of neighborhoods of x. Now suppose that we start with a neighborhood space $(X, \mathfrak{N})$. We define open set in $(X, \mathfrak{N})$ by Definition 3.5, thus obtaining a topological space $(X, \mathfrak{I})$. In the topological space $(X, \mathfrak{I})$ we define neighborhood by Definition 2.2 to obtain a neighborhood space $(X, \mathfrak{N}')$. Under these circumstances, if $N \in \mathfrak{N}_x$, by Lemma 3.7, N contains an open set O containing x, so that by Definition 2.2, N is a neighborhood of x in $(X, \mathfrak{I})$, or $N \in \mathfrak{N}'_x$. Conversely, if $N \in \mathfrak{N}'_x$, then by Definition 2.2, N contains a set $O \in \mathfrak{I}$, and $x \in O$. Since $O \in \mathfrak{I}$, O is open in the neighborhood space $(X, \mathfrak{N})$ and so by Lemma 3.7, N is a neighborhood of x. Thus, for each $x \in X$, $\mathfrak{N}_x = \mathfrak{N}'_x$, and the two neighborhood spaces are the same.

Collecting together the results on the correspondence between topological spaces and neighborhood spaces, we have:

THEOREM 3.8 Let neighborhood in a topological space be defined by Definition 2.2 and open set in a neighborhood space be defined by Definition 3.5. Then the neighborhoods of a topological space $(X, \mathfrak{I})$ give rise to a neighborhood space $(X, \mathfrak{N}) = \mathfrak{A}(X, \mathfrak{I})$ and the open sets of a neighborhood space $(Y, \mathfrak{N}')$ give rise to a topological space $(Y, \mathfrak{I}') = \mathfrak{A}'(Y, \mathfrak{N}')$. Furthermore, for each topological space $(X, \mathfrak{I})$,

$$(X, \mathfrak{I}) = \mathfrak{A}'(\mathfrak{A}(X, \mathfrak{I})),$$

and for each neighborhood space $(X, \mathfrak{N})$,

$$(X, \mathfrak{N}) = \mathfrak{A}(\mathfrak{A}'(X, \mathfrak{N})),$$

thus establishing a one-one correspondence between the collection of all topological spaces and the collection of all neighborhood spaces.

Theorem 3.8 justifies the specification of a topological space by defining for a given set X what subsets of X are to be the neighborhoods of a point $x \in X$; that is, by specifying the corresponding neighborhood space. For example, let X be the set of positive integers. Given a point $n \in X$ and a subset U of X, let us call U a neighborhood of n if for each integer $m \geqq n$, $m \in U$. We must then verify that these neighborhoods satisfy conditions $N1$–$N5$ so that we have a neighborhood space and consequently a topological space. The reader should verify that this corresponding topological space is the one described in Example 7 of Section 2.

EXERCISES

1. Given a real number x, call a subset N of R a neighborhood of x if $y \geqq x$ implies $y \in N$. Prove that this definition of neighborhood yields a neighborhood space. Describe the corresponding topological space.
2. Given a real number x, call a subset N of R a neighborhood of x if N contains the closed interval $[x, x + 1]$. Prove that the neighborhoods so defined satisfy $N1$–$N4$, but not $N5$. Use the Definition 3.5 of open set anyway, and determine which subsets of R will be open.
3. In a neighborhood space, a collection $\mathfrak{B}_x$ of neighborhoods of a point $x \in X$ is called a *basis for the complete system of neighborhoods at x*, or simply a *basis for the neighborhoods at x*, if, for each neighborhood N of x, there is a neighborhood $U \in \mathfrak{B}_x$ such that $U \subset N$.

   Prove that if for each point $x \in X$, $\mathfrak{B}_x$ is a basis for the neighborhoods at x, then:

*BN*1. For each $x \in X$, $\mathcal{B}_x \neq \emptyset$;

*BN*2. For each $x \in X$ and $U \in \mathcal{B}_x$, $x \in U$;

*BN*3. For each $x \in X$ and $U, V \in \mathcal{B}_x$, $U \cap V$ contains an element $W \in \mathcal{B}_x$;

*BN*4. For each $x \in X$ and $U \in \mathcal{B}_x$, there is an $O \subset U$ such that $x \in O$ and for each $y \in O$, O contains an element $V_y \in \mathcal{B}_y$.

4. Define a *basic neighborhood space* to be a set X, such that for each $x \in X$ a collection $\mathcal{B}_x$ of subsets of X satisfies the conditions *BN*1–*BN*4 of Problem 3. In a basic neighborhood space $(X, \mathcal{B})$ define a subset N of X to be a neighborhood of a point $x \in X$, if $N \supset U$ for some $U \in \mathcal{B}_x$. Prove that the neighborhoods of a basic neighborhood space yield a neighborhood space. (Thus a topological space may be constructed by specifying for each point x a basis $\mathcal{B}_x$ of the neighborhoods at x satisfying *BN*1–*BN*4.) The correspondence between basic neighborhood spaces and neighborhood spaces is many-one, since there are many different bases for the neighborhoods at a point in a neighborhood space. However, prove that if $(X, \mathcal{B})$ and $(X, \mathcal{B}')$ are two basic neighborhood spaces, then they give rise to the same neighborhood space if and only if for each point $x \in X$ we have

   (i) given $U \in \mathcal{B}_x$, there is a $U' \in \mathcal{B}'_x$ with $U' \subset U$, and
   (ii) given $V' \in \mathcal{B}'_x$, there is a $V \in \mathcal{B}_x$ with $V \subset V'$.

   Also prove that starting from a given neighborhood space $(X, \mathfrak{N})$, if for each $x \in X$, $\mathcal{B}_x$ is a basis for the neighborhoods at x, then the neighborhood space that arises from the basic neighborhood space $(X, \mathcal{B})$ is $(X, \mathfrak{N})$.

4 CLOSURE, INTERIOR, BOUNDARY

In a metric space, given a point x and a subset A, we can say that there are points of A arbitrarily close to x if $d(x, A) = 0$. In a topological space, we can also find a characterization of "arbitrary closeness." To indicate the proper translation from metric spaces to topological spaces of this concept, let us first prove:

Lemma 4.1 In a metric space (X, d), for a given point x and a given subset A, $d(x, A) = 0$ if and only if each neighborhood N of x contains a point of A.

Proof. First, suppose that each neighborhood N of x contains a point of A. In particular, for each $\varepsilon > 0$, there is a point of A in $B(x; \varepsilon)$. Thus $\text{g.l.b.}_{a \in A} \{d(x, a)\} < \varepsilon$ for each $\varepsilon > 0$ and consequently $d(x, A) = \text{g.l.b.}_{a \in A} \{d(x, a)\} = 0$. Conversely, suppose that there is a neighborhood N of x that does not contain a point of A. Since N is a neighborhood of x in a metric space, there is an $\varepsilon > 0$ such that $B(x; \varepsilon) \subset N$. It follows that $a \in A$ implies that $d(x, a) \geqq \varepsilon$. Thus $d(x, A) \geqq \varepsilon$.

We shall, therefore, in a topological space, say that the points of a subset A are arbitrarily close to a given point x, if each neighborhood of x contains a point of A. Given a subset A, the collection of points that are arbitrarily close to A is called the *closure* of A.

Definition 4.2 Let A be a subset of a topological space. A point x is said to be *in the closure of* A if, for each neighborhood N of x, $N \cap A \neq \emptyset$. The closure of A is denoted by $\overline{A}$.

The purpose of the next two lemmas is to provide a description of the closure of a subset in terms of closed sets.

Lemma 4.3 Given a subset A of a topological space and a closed set F containing A, $\overline{A} \subset F$.

Proof. Suppose $x \notin F$, then x is in the open set $C(F)$. Also, $F \supset A$ implies $C(F) \subset C(A)$. Thus, $C(F) \cap A = \emptyset$. Since $C(F)$ is a neighborhood of x, $x \notin \overline{A}$. We have thus shown that $C(F) \subset C(\overline{A})$ or $\overline{A} \subset F$.

Lemma 4.4 Given a subset A of a topological space and a point $x \notin \overline{A}$, then $x \notin F$ for some closed set F containing A.

Proof. If $x \notin \overline{A}$, then there is a neighborhood and hence an open set O containing x such that $O \cap A = \emptyset$. Let $F = C(O)$. Then F is closed and $F = C(O) \supset A$. But $x \in O$ and therefore $x \notin F$.

Combining these two lemmas, we obtain:

THEOREM 4.5 Given a subset A of a topological space, $\overline{A} = \bigcap_{\alpha \in I} F_\alpha$, where $\{F_\alpha\}_{\alpha \in I}$ is the family of all closed sets containing A.

Proof. By Lemma 4.3, $\overline{A} \subset \bigcap_{\alpha \in I} F_\alpha$ since $\overline{A} \subset F_\alpha$ for each $\alpha \in I$. By Lemma 4.4, $x \in F_\alpha$ for each $\alpha \in I$ implies that $x \in \overline{A}$, or $\bigcap_{\alpha \in I} F_\alpha \subset \overline{A}$. Thus, $\overline{A} = \bigcap_{\alpha \in I} F_\alpha$.

Frequently, in introducing the concept of closure of a subset, the characterization of closure given by Theorem 4.5 is used as a definition and the statement embodied in our Definition, 4.2, is then proved as a theorem. Another possible description of the closure $\bar{A}$ of a subset A is the characterization of $\bar{A}$ as the smallest closed set containing A. For $\bar{A}$ is contained in each closed set containing A, while $\bar{A}$, being the intersection of closed sets, is itself a closed set.

Theorem 4.5 is the characterization of closure in terms of closed sets. The next theorem characterizes closed sets in terms of closure.

THEOREM 4.6 A is closed if and only if $A = \overline{A}$.

Proof. We have just seen that $\overline{A}$ is closed, so if $A = \overline{A}$, then A is closed. Conversely, suppose A is closed. In this event A itself is a closed set containing A, so, therefore, $\overline{A} \subset A$. On the other hand, for an arbitrary subset A, we have $A \subset \overline{A}$, for if $x \in A$, then each neighborhood N of x contains a point of A; namely x itself. Thus, if A is closed, $A = \overline{A}$.

The act of taking the closure of a set associates to each subset A of a topological space a new subset $\bar{A}$. This correspondence or operation on the subsets satisfies the following five properties:

Theorem 4.7 In a topological space $(X, \mathfrak{I})$,

*CL*1. $\overline{\emptyset} = \emptyset$;
*CL*2. $\overline{X} = X$;
*CL*3. For each subset A of X, $A \subset \overline{A}$;
*CL*4. For each pair of subsets A, B of X, $\overline{A \cup B} = \overline{A} \cup \overline{B}$;
*CL*5. For each subset A of X, $\overline{\overline{A}} = \overline{A}$.

Proof. The property *CL*3 has been established during the proof of Theorem 4.6. Note that *CL*2 follows from *CL*3. *CL*1 is true, for given a point $x \in X$ and a neighborhood N of x, $N \cap \emptyset = \emptyset$; thus there are no points in $\overline{\emptyset}$. To prove *CL*5 we note that $\overline{A}$ is closed, so, applying Theorem 4.6 to $\overline{A}$ we have $\overline{\overline{A}} = \overline{A}$. It remains for us to prove *CL*4. Suppose $x \in \overline{A}$, then each neighborhood N of x contains points of A and hence points of $A \cup B$. Thus $\overline{A} \subset \overline{A \cup B}$. Similarly, $\overline{B} \subset \overline{A \cup B}$, and, consequently, $\overline{A} \cup \overline{B} \subset \overline{A \cup B}$. On the other hand, $A \subset \overline{A}$ and $B \subset \overline{B}$, so $A \cup B \subset \overline{A} \cup \overline{B}$. Thus, $\overline{A} \cup \overline{B}$ is a closed set containing $A \cup B$, whence $\overline{A \cup B} \subset \overline{A} \cup \overline{B}$.

One may use the properties *CL*1–*CL*5 as a set of axioms for what we will call a *closure space* and then prove that there is a "natural" one-one correspondence between the collection of topological spaces and the collection of closure spaces. An outline of how this might be done is given in Problem 11 at the end of this section.

In a topological space, we have seen that the closure of a subset A is the smallest closed set containing A. Another significant subset associated with A is the "interior" of A, which, as we shall see, is the largest open set contained in A.

Definition 4.8 Given a subset A of a topological space, a point x is said to be *in the interior of* A if A is a neighborhood of x. The interior of A is denoted by $\text{Int}(A)$.

Lemma 4.9 Given a subset A of a topological space and an open set O contained in A, $O \subset \text{Int}(A)$.

Proof. If $x \in O$, then A is a neighborhood of x, since O is open and $O \subset A$. Thus $x \in \text{Int}(A)$ and $O \subset \text{Int}(A)$.

LEMMA 4.10 Given a subset A of a topological space, if $x \in \text{Int}(A)$, then $x \in O$ for some open set $O \subset A$.

Proof. If $x \in \text{Int}(A)$, then A is a neighborhood of x, whence A contains an open set O containing x.

In much the same manner in which Lemmas 4.3 and 4.4 combine to yield Theorem 4.5, Lemmas 4.9 and 4.10 combine to yield:

THEOREM 4.11 Given a subset A of a topological space,

$$\text{Int}(A) = \bigcup_{\alpha \in I} O_\alpha,$$

where $\{O_\alpha\}_{\alpha \in I}$ is the family of all open sets contained in A.

Thus, Int (A), being the union of open sets, is itself open, and is the largest open set contained in A. Furthermore, if $\{O_\alpha\}_{\alpha \in I}$ is the family of open sets contained in a given set A, then $\{C(O_\alpha)\}_{\alpha \in I}$ is the family of closed sets containing $C(A)$. Thus:

THEOREM 4.12 $C(\text{Int}(A)) = \overline{C(A)}$.

COROLLARY 4.13 $\text{Int}(A) = C(\overline{C(A)})$, $C(\bar{A}) = \text{Int}(C(A))$.

For a given subset A, the set of points that are arbitrarily close to both A and $C(A)$ is called the "boundary" of A.

DEFINITION 4.14 Given a subset A of a topological space, a point x is said to be *in the boundary of* A if x is in both the closure of A and the closure of the complement of A. The boundary of A is denoted by Bdry (A).

Thus, Bdry $(A) = \bar{A} \cap \overline{C(A)}$. It follows that A and $C(A)$

have the same boundary, for Bdry $C(A) = \overline{C(A)} \cap \overline{C(C(A))} = \overline{C(A)} \cap \bar{A}$. In terms of the definition of the closure of a set, we have the statement that a point x is in the boundary of a set A if and only if each neighborhood N of x contains both points of A and points of the complement of A. Since the boundary of A is the intersection of two closed sets:

COROLLARY 4.15 For each subset A, Bdry (A) is closed.

EXERCISES

1. A family $\{A_\alpha\}_{\alpha \in I}$ of subsets is said to be *mutually disjoint* if for each distinct pair β, γ of indices $A_\beta \cap A_\gamma = \emptyset$. Prove that for each subset A of a topological space $(X, \mathfrak{J})$, the three sets Int (A), Bdry (A), and Int $(C(A))$ are mutually disjoint and that $X =$ Int $(A) \cup$ Bdry $(A) \cup$ Int $(C(A))$.
2. In a metric space (X, d), prove that for each subset A:
   (*a*) $x \in \bar{A}$ if and only if $d(x, A) = 0$;
   (*b*) $x \in$ Int (A) if and only if $d(x, C(A)) > 0$;
   (*c*) $x \in$ Bdry (A) if and only if
   $$d(x, A) = 0 \quad \text{and} \quad d(x, C(A)) = 0.$$
3. In the real line, prove that the boundary of the open interval (a, b) and the boundary of the closed interval $[a, b]$ is $\{a, b\}$.
4. In R^n with the usual topology, let A be the set of points $x = (x_1, x_2, \ldots, x_n)$ such that $x_1^2 + x_2^2 + \ldots + x_n^2 \leqq 1$. Prove that Bdry (A) is the $(n - 1)$-dimensional sphere S^{n-1}, i.e., $x \in$ Bdry (A) if and only if $x_1^2 + x_2^2 + \ldots + x_n^2 = 1$.
5. In R^{n+1} with the usual topology, let A be the set of points $x = (x_1, x_2, \ldots, x_{n+1})$ such that $x_{n+1} = 0$. Prove that Int $(A) = \emptyset$, Bdry $(A) = A$, $\bar{A} = A$.
6. In a topological space, each of the terms *open set, closed set, neighborhood, closure of a set, interior of a set, boundary of a set,* may be characterized by any other one of these terms. Construct a table containing the thirty such possible definitions or theorems in which, for example, the entry in the row labelled interior and the column

labelled open set is the characterization of interior in terms of open sets (Theorem 4.17), etc.

7. Let A be a subset of a topological space. Prove that Bdry $(A) = \emptyset$ if and only if A is open and closed.
8. A subset A of a topological space $(X, \mathfrak{I})$ is said to be *dense in* X if $\overline{A} = X$. Prove that if for each open set O we have $A \cap O \neq \emptyset$, then A is dense in X.
9. The "rational density theorem" for the real line states that between any two real numbers there lies a rational number. Use the rational density theorem to prove that the rational numbers are dense in the real line.
10. The "Archimedean principle" for the real line states that if $c, d > 0$ then there is a positive integer N such that $Nc > d$. Prove the Archimedean principle for the real line and use this principle to prove the rational density theorem for the real line.
11. Let a *closure space* be defined as a set X together with a correspondence which associates to each subset A of X a subset $\overline{A}$ of X satisfying the five conditions $CL1$–$CL5$ of Theorem 4.7. Prove that in a closure space, $A \subset B$ implies $\overline{A} \subset \overline{B}$. Define A to be closed if $A = \overline{A}$. Prove that the empty set and the whole space are closed. Also that a finite union of closed sets is closed and an arbitrary intersection of closed sets is closed. Prove that for each subset A of X, $\overline{A} = \bigcap_{\alpha \in I} F_\alpha$, where $\{F_\alpha\}_{\alpha \in I}$ is the family of all closed sets containing A. Now prove that there is a one-one correspondence between the collection of topological spaces and the collection of closure spaces.
12. Prove that $\overline{A} = A \cup$ Bdry (A).
13. Let A be a subset of a topological space. Prove that A is closed if and only if Bdry $(A) \subset A$, and that A is open if and only if Bdry $(A) \subset C(A)$.

5 FUNCTIONS, CONTINUITY, HOMEOMORPHISM

Definition 5.1 A *function* f from a topological space $(X, \mathfrak{I})$ to a topological space $(Y, \mathfrak{I}')$ is a function $f: X \to Y$.

If f is a function from a topological space $(X, \mathfrak{I})$ to a topological space $(Y, \mathfrak{I}')$ we shall write $f:(X, \mathfrak{I}) \to (Y, \mathfrak{I}')$. In the event that the topologies on X and Y need not be explicitly mentioned, we may abbreviate this notation by $f:X \to Y$ or simply f.

Definition 5.2 A function $f:(X, \mathfrak{I}) \to (Y, \mathfrak{I}')$ is said to be *continuous at a point* $a \in X$ if for each neighborhood N of $f(a)$, $f^{-1}(N)$ is a neighborhood of a. f is said to be *continuous* if f is continuous at each point of X.

Let (X, d) and (Y, d') be metric spaces and let their associated topological spaces be $(X, \mathfrak{I})$ and $(Y, \mathfrak{I}')$ respectively. Given a function f from the first metric space to the second, we also have a function, which we still denote by f, from the first topological space to the second. Our definition of continuity has been formulated so that for each point $a \in X$, the function $f:(X, d) \to (Y, d')$ is continuous at a if and only if $f:(X, \mathfrak{I}) \to (Y, \mathfrak{I}')$ is continuous at a.

Theorem 5.3 A function $f:(X, \mathfrak{I}) \to (Y, \mathfrak{I}')$ is continuous if and only if for each open subset O of Y, $f^{-1}(O)$ is an open subset of X.

Proof. First, suppose that f is continuous and that O is an open subset of Y. For each $a \in f^{-1}(O)$, O is a neighborhood of $f(a)$, therefore $f^{-1}(O)$ is a neighborhood of a. Since $f^{-1}(O)$ is a neighborhood of each of its points, $f^{-1}(O)$ is an open subset of X. Conversely, suppose that for each open subset O of Y, $f^{-1}(O)$ is an open subset of X. Let $a \in X$ and a neighborhood N of $f(a)$ be given. N contains an open set O containing $f(a)$, so by our hypothesis, $f^{-1}(N)$ contains the open set $f^{-1}(O)$ containing a. Thus, $f^{-1}(N)$ is a neighborhood of a and f is continuous at a. Since a was arbitrary, f is continuous.

For any set X, given a collection $\mathbf{E}$ of subsets of X, let $C'(\mathbf{E})$ denote the collection of subsets of X which are complements of members of **E**. Also given $f: X \to Y$ and a collection **E** of subsets of Y, let $f^{-1}(\mathbf{E})$ denote the collection of subsets of X of the form $f^{-1}(E)$ for some $E \in \mathbf{E}$. Theorem 5.3 states that $f: (X, \mathfrak{I}) \to (Y, \mathfrak{I}')$ is continuous if and only if $f^{-1}(\mathfrak{I}') \subset \mathfrak{I}$. Let $\mathfrak{F} = C'(\mathfrak{I})$ and $\mathfrak{F}' = C'(\mathfrak{I}')$ be the closed subsets of X and Y respectively. For $F \in \mathfrak{F}'$, $f^{-1}(C(F)) = C(f^{-1}(F))$ so that $f^{-1}(\mathfrak{F}') = C'(f^{-1}(\mathfrak{I}'))$. Thus $f^{-1}(\mathfrak{I}') \subset \mathfrak{I}$ is equivalent to $f^{-1}(\mathfrak{F}') \subset \mathfrak{F}$ and we obtain:

THEOREM 5.4 A function $f: (X, \mathfrak{I}) \to (Y, \mathfrak{I}')$ is continuous if and only if for each closed subset F of Y, $f^{-1}(F)$ is a closed subset of X.

It is important to remember that Theorem 5.3 says that a function f is continuous if and only if the *inverse* image of each open set is open. This characterization of continuity should not be confused with another property that a function may or may not possess, the property that the image of each open set is an open set (such functions are called *open mappings*). There are many situations in which a function $f: (X, \mathfrak{I}) \to (Y, \mathfrak{I}')$ has the property that for each open subset A of X, the set $f(A)$ is an open subset of Y, and yet f is *not* continuous. For example, let Y be a set containing two distinct elements a and b and let each subset of Y be an open set. Let R be the real line and define $f: R \to Y$ by $f(x) = a$ for $x \geqq 0$ and $f(x) = b$ for $x < 0$. Every subset of Y is open so in particular for each open subset U of R, $f(U)$ is an open subset of Y. On the other hand $\{a\}$ is an open subset of Y but $f^{-1}(\{a\})$, the set of non-negative real numbers, is not an open subset of the reals.

THEOREM 5.5 $f: (X, \mathfrak{I}) \to (Y, \mathfrak{I}')$ is continuous if and only if for each subset A of X, $f(\overline{A}) \subset \overline{f(A)}$.

Proof. First suppose that f is continuous. Given a subset A of X, $f(A) \subset \overline{f(A)}$, whence $A \subset f^{-1}(f(A)) \subset f^{-1}(\overline{f(A)})$. The set $f^{-1}(\overline{f(A)})$ is closed so $\overline{A} \subset f^{-1}(\overline{f(A)})$. Thus $f(\overline{A}) \subset \overline{f(A)}$. Conversely, suppose that for each subset A of X, $f(\overline{A}) \subset \overline{f(A)}$. Let F be a closed subset of Y. Then $f(\overline{f^{-1}(F)}) \subset \overline{f(f^{-1}(F))} \subset \overline{F} = F$. Thus

$\overline{f^{-1}(F)} \subset f^{-1}(F)$. Since it is always the case that $f^{-1}(F) \subset \overline{f^{-1}(F)}$ we have $f^{-1}(F) = \overline{f^{-1}(F)}$; consequently $f^{-1}(F)$ is closed and f is continuous.

THEOREM 5.6 Let $f:(X, \mathfrak{I}) \to (Y, \mathfrak{I}')$ be continuous at a point $a \in X$ and let $g:(Y, \mathfrak{I}') \to (Z, \mathfrak{I}'')$ be continuous at $f(a)$. Then the composite function $gf:(X, \mathfrak{I}) \to (Z, \mathfrak{I}'')$ is continuous at a.

Proof. Let N be a neighborhood of $(gf)(a) = g(f(a))$. Then $(gf)^{-1}(N) = f^{-1}(g^{-1}(N))$. But $g^{-1}(N)$ is a neighborhood of $f(a)$, since g is continuous at $f(a)$, and therefore $f^{-1}(g^{-1}(N))$ is a neighborhood of a, since f is continuous at a.

The equivalence relation that is appropriate to topological spaces is called *homeomorphism.*

DEFINITION 5.7 Topological spaces $(X, \mathfrak{I})$ and $(Y, \mathfrak{I}')$ are called *homeomorphic* if there exist inverse functions $f:X \to Y$ and $g:Y \to X$ such that f and g are continuous. In this event the functions f and g are said to be *homeomorphisms* and we say that f and g *define a homeomorphism between* $(X, \mathfrak{I})$ *and* $(Y, \mathfrak{I}')$.

The following easily verified corollary to this definition indicates that homeomorphism is the translation from metric spaces to topological spaces of the concept of topological equivalence.

COROLLARY 5.8 Let (X, d) and (Y, d') be metric spaces. Let $(X, \mathfrak{I})$ and $(Y, \mathfrak{I}')$ be the topological spaces associated with (X, d) and (Y, d') respectively. Then the metric spaces (X, d) and (Y, d') are topologically equivalent if and only if the topological spaces $(X, \mathfrak{I})$ and $(Y, \mathfrak{I}')$ are homeomorphic.

THEOREM 5.9 A necessary and sufficient condition that two topological spaces $(X, \mathfrak{I})$ and $(Y, \mathfrak{I}')$ be homeomorphic is that there exist a function $f: X \to Y$ such that:

1. f is one-one;
2. f is onto;
3. A subset O of X is open if and only if $f(O)$ is open.

Proof. Suppose that $(X, \mathfrak{I})$ and $(Y, \mathfrak{I}')$ are homeomorphic. Let the homeomorphism be defined by inverse functions $f: X \to Y$ and $g: Y \to X$. f is invertible and consequently one-one and onto. Furthermore, given an open set O in X, the set $f(O) = g^{-1}(O)$ is open in Y, since g is continuous. On the other hand, if $f(O) = O'$ is an open subset of Y, then $O = f^{-1}(O')$ is open in X.

Now, suppose that a function $f: X \to Y$ with the prescribed properties exists. Then f is invertible and we define $g: Y \to X$ by $g(b) = a$ if $f(a) = b$, so that f and g are inverse functions. If O is an open subset of X, then $f(O) = g^{-1}(O)$ is open in Y, so that g is continuous. Also, if O' is an open subset of Y, then $f^{-1}(O') = O$ is an open subset of X and f is continuous.

EXERCISES

1. Let a function $f: X \to Y$ be given. Prove that $f: (X, 2^X) \to (Y, \mathfrak{I}')$ is always continuous, as is $f: (X, \mathfrak{I}) \to (Y, \{\emptyset, Y\})$, where $\mathfrak{I}'$ is any topology on Y and $\mathfrak{I}$ is any topology on X.
2. Prove that a function $f: (X, \mathfrak{I}) \to (Y, \mathfrak{I}')$ is a homeomorphism if and only if
   (i) f is one-one;
   (ii) f is onto;
   (iii) For each point $x \in X$ and each subset N of X, N is a neighborhood of x if and only if $f(N)$ is a neighborhood of $f(x)$.
3. Let $f: (X, \mathfrak{I}) \to (Y, \mathfrak{I}')$ be a homeomorphism. Let a third topological space $(Z, \mathfrak{I}'')$ and a function $h: (Y, \mathfrak{I}') \to (Z, \mathfrak{I}'')$ be given. Prove that h is continuous if and only if hf is continuous. Let another function $k: (Z, \mathfrak{I}'') \to (X, \mathfrak{I})$ be given. Prove that k is continuous if and only if fk is continuous.
4. Let R be the real line. Prove that the function $f: R \to R$

defined by $f(x) = \sin x$ is continuous. $\left[\textit{Hint: } |\sin a - \sin x| = 2\left|\sin \frac{a-x}{2} \cos \frac{a+x}{2}\right| \text{ and } \left|\sin \frac{a-x}{2}\right| \leqq \left|\frac{a-x}{2}\right|.\right]$ Find an open interval (a, b) such that $f((a, b))$ is *not* an open interval.

6 SUBSPACES

DEFINITION 6.1 Let $(X, \mathfrak{I})$ and $(Y, \mathfrak{I}')$ be topological spaces. The topological space Y is called a *subspace* of the topological space X if $Y \subset X$ and if the open subsets of Y are precisely the subsets O' of the form

$$O' = O \cap Y$$

for some open subset O of X.

In the event that Y is a subspace of X, we may say that each open subset O' of Y is the restriction to Y of an open subset O of X. A subset O' that is open in Y is often called *relatively open in Y* or simply *relatively open*. A subset O of X that is open in X and is contained in Y is necessarily relatively open in Y, but the relatively open subsets of Y are in general not open in X.

We shall now prove that there are as many subspaces of a topological space X as there are non-empty subsets Y of X.

PROPOSITION 6.2 Let $(X, \mathfrak{I})$ be a topological space and let Y be a subset of X. Define the collection $\mathfrak{I}'$ of subsets of Y as the collection of subsets O' of Y of the form

$$O' = O \cap Y,$$

where $O \in \mathfrak{I}$. Then $(Y, \mathfrak{I}')$ is a topological space and therefore a subspace of $(X, \mathfrak{I})$ provided $Y \neq \emptyset$.

Proof. We must prove that $\mathfrak{I}'$ is a topology. $\emptyset = \emptyset \cap Y$ and $Y = X \cap Y$ are in $\mathfrak{I}$. Suppose $O_1', O_2', \ldots, O_n' \in \mathfrak{I}'$, so that for $i = 1, 2, \ldots, n$, $O_i' = O_i \cap Y$ for some $O_i \in \mathfrak{I}$. Then

$$O_1' \cap O_2' \cap \ldots \cap O_n' = (O_1 \cap O_2 \cap \ldots \cap O_n) \cap Y$$

is in $\mathfrak{J}'$, since $O_1 \cap O_2 \cap \ldots \cap O_n$ is open in X. Finally, suppose that for each $\alpha \in I$, $O_\alpha' \in \mathfrak{J}'$. Thus, for each $\alpha \in I$, $O_\alpha' = O_\alpha \cap Y$ for some $O_\alpha \in \mathfrak{J}$. But $\bigcup_{\alpha \in I} O_\alpha' = \bigcup_{\alpha \in I} (O_\alpha \cap Y) = (\bigcup_{\alpha \in I} O_\alpha) \cap Y$ is in $\mathfrak{J}'$, since $\bigcup_{\alpha \in I} O_\alpha$ is open in X.

Given a subset Y of a topological space $(X, \mathfrak{J})$, the topology $\mathfrak{J}'$ of Y described in the above proposition is said to be *induced* by the topology $\mathfrak{J}$ on X and is called the *relative topology* on Y. The neighborhoods in this relative topology on Y are called *neighborhoods in Y* or *relative neighborhoods*. The following result states that the neighborhoods in Y are the restrictions of the neighborhoods in X.

THEOREM 6.3 Let Y be a subspace of a topological space X and let $a \in Y$. Then a subset N' of Y is a relative neighborhood of a if and only if

$$N' = N \cap Y,$$

where N is a neighborhood of a in X.

Proof. First suppose N' is a relative neighborhood of a. Then N' contains a relatively open set O', which contains a. Let $O' = O \cap Y$, where O is an open subset of X. Then $N = N' \cup O$ is a neighborhood of a in X and $N \cap Y = (N' \cup O) \cap Y = N' \cup (O \cap Y) = N'$. Conversely, suppose $N' = N \cap Y$, where N is a neighborhood of a in X. Then N contains an open set O containing a and hence N' contains the relatively open set $O' = O \cap Y$ containing a. Thus N' is a relative neighborhood of a.

EXAMPLE 1 The closed interval $[a, b]$ of the real line with induced topology is a subspace of the real line. A relative neighborhood of the point a is any subset N of $[a, b]$ that contains a half-open interval $[a, c)$, where $a < c$ and $[a, c)$ is the set of all real numbers x such that $a \leqq x < c$. Similarly, a relative neighborhood of the point b is any subset M of

$[a, b]$ that contains a half-open interval $(c, b]$, where $c < b$ and $(c, b]$ is the set of all real numbers x such that $c < x \leqq b$. If d is such that $a < d < b$, then a relative neighborhood of d is any subset U of $[a, b]$ that is a neighborhood of d in the real line R.

EXAMPLE 2 Let A be the subset of R^{n+1} consisting of all points $x = (x_1, x_2, \ldots, x_{n+1})$ such that $x_{n+1} = 0$. Let R^{n+1} have the usual topology and let A have the induced topology so that A is a subspace of R^{n+1}. The topological space A is homeomorphic to R^n. To prove this fact we shall use the result that the relationship of subspace is "preserved" in passing from metric spaces to topological spaces.

LEMMA 6.4 Let (X, d) be a metric space and let (Y, d') be a subspace of (X, d). If $(X, \mathfrak{I})$ is the topological space associated with (X, d) and $(Y, \mathfrak{I}')$ is the topological space associated with (Y, d'), then $(Y, \mathfrak{I}')$ is a subspace of $(X, \mathfrak{I})$.

Proof. Since d' is the restriction of d, an open ball in (Y, d') is the restriction of an open ball in (X, d) to Y. Consequently a subset O' of Y is open in Y if and only if, for each $y \in O'$, there is an $\varepsilon_y > 0$ such that $B(y; \varepsilon_y) \cap Y \subset O'$. Let $O = \bigcup_{y \in O'} B(y; \varepsilon_y)$. Then O is open in X and $O' = O \cap Y$. Thus $O' \in \mathfrak{I}'$. Conversely, if $O' \in \mathfrak{I}'$, then $O' = O \cap Y$ for some $O \in \mathfrak{I}$. For each $y \in O'$ we have $y \in O$, and O is open, so there is an ε_y such that $B(y; \varepsilon_y) \subset O$. It follows that $B(y; \varepsilon_y) \cap Y \subset O'$, and hence O' is open in (Y, d').

Returning to our example, we define $f: R^n \to A$ by setting $f(x_1, x_2, \ldots, x_n) = (x_1, x_2, \ldots, x_n, 0)$. It is easily verified that f is one-one, onto, and that the inverse of f is the function $g: A \to R^n$ defined by $g(x_1, x_2, \ldots, x_n, 0) = (x_1, x_2, \ldots, x_n)$. If we first consider f and g as functions defined on the metric spaces (R^n, d) and (A, d'), where (A, d') is a subspace of (R^{n+1}, d), then clearly f and g are continuous. It follows that f and g are continuous functions defined on the topological spaces R^n and A, where A is considered as a subspace of R^{n+1}, and that they therefore define a homeomorphism.

Given a subspace $(Y, \mathfrak{J}')$ of a topological space $(X, \mathfrak{J})$, the closed subsets of the topological space $(Y, \mathfrak{J}')$ are called *relatively closed in Y* or simply *relatively closed.* Again, the relatively closed subsets are the restriction to Y of the closed subsets of X.

THEOREM 6.5 Let $(Y, \mathfrak{J}')$ be a subspace of the topological space $(X, \mathfrak{J})$. A subset F' of Y is relatively closed in Y if and only if

$$F' = F \cap Y,$$

for some closed subset F of X.

Proof. First, suppose F' is relatively closed. Then $C_Y(F')$ is relatively open. Thus, $C_Y(F') = O \cap Y$, where O is open in X. But then $F' = C_Y(O \cap Y) = C_Y(O) = C_X(O) \cap Y$, where $C_X(O)$ is a closed subset of X. Conversely, suppose $F' = F \cap Y$, where F is a closed subset of X. Then, $C_Y(F') = C_X(F) \cap Y$; hence $C_Y(F')$ is relatively open in Y and therefore F' is relatively closed.

EXAMPLE 3 Let $a < b < c < d$. Let $Y = [a, b] \cup (c, d)$ be considered as a subspace of the real line. Then the subset $[a, b]$ of Y is both relatively open and relatively closed. To prove this fact we note that $[a, b] = [a, b] \cap Y$ so that $[a, b]$ is relatively closed, whereas for $0 < \varepsilon < c - b$, $[a, b] = (a - \varepsilon, b + \varepsilon) \cap Y$ so that $[a, b]$ is relatively open. Since (c, d) is the complement in Y of a relatively open and relatively closed subset of Y, (c, d) is also relatively open and relatively closed in Y.

THEOREM 6.6 Let the topological space Y be a subspace of the topological space X. Then the inclusion mapping $i: Y \to X$ is continuous.

Proof. For each subset A of X, $i^{-1}(A) = A \cap Y$. Thus, if O is an open subset of X, $i^{-1}(O) = O \cap Y$ is a relatively open subset of Y.

DEFINITION 6.7 Let $\mathfrak{J}_1$ and $\mathfrak{J}_2$ be two topologies on a set Y. The topology $\mathfrak{J}_1$ is said to be *weaker* than $\mathfrak{J}_2$ if $\mathfrak{J}_1 \subset \mathfrak{J}_2$.

If Y is a subset of a topological space $(X, \mathfrak{I})$ then the relative topology $\mathfrak{I}'$ on Y is the weakest topology such that the inclusion map $i: Y \to X$ is continuous, for if $\mathfrak{I}_1$ is another topology on Y such that $i:(Y, \mathfrak{I}_1) \to (X, \mathfrak{I})$ is continuous then given $O' \in \mathfrak{I}'$, $O' = i^{-1}(O)$ with $O \in \mathfrak{I}$. Thus $O' \in \mathfrak{I}_1$ and $\mathfrak{I}' \subset \mathfrak{I}_1$.

Let X and Y be topological spaces and $f: Y \to X$ be a function which is not necessarily continuous. The function f induces a function $f': Y \to f(Y)$ which agrees with f and is onto. Viewing $f(Y)$ as a subspace of X we have:

LEMMA 6.8 $f: Y \to X$ is continuous if and only if $f': Y \to f(Y)$ is continuous.

Proof. Since the inclusion map $i: f(Y) \to X$ is continuous, the continuity of f' yields the continuity of $f = if'$. Conversely, if O' is a relatively open set in $f(Y)$, then $O' = O \cap f(Y)$, where O is open in X. If f is continuous $f^{-1}(O) = f'^{-1}(O')$ is open in Y and f' is continuous.

EXERCISES

1. If Y is a subspace of X and Z is a subspace of Y, then Z is a subspace of X.
2. Let O be an open subset of a topological space X. Prove that a subset A of O is relatively open in O if and only if it is an open subset of X.
3. Let F be a closed subset of a topological space X. Prove that a subset A of F is relatively closed in F if and only if it is a closed subset of X.
4. Prove that a subspace of a Hausdorff space is a Hausdorff space.
5. Prove that a subspace of a metrizable space is a metrizable space.
6. Prove that an open interval (a, b) considered as a subspace of the real line is homeomorphic to the real line.
7. Let Y be a subspace of X and let A be a subset of Y. Denote by $\text{Int}_X(A)$ the interior of A in the topological space X and by $\text{Int}_Y(A)$ the interior of A in the topological space Y. Prove that $\text{Int}_X(A) \subset \text{Int}_Y(A)$. Illustrate by an example the fact that in

general $\text{Int}_X(A) \neq \text{Int}_Y(A)$.

8. Let Y be a subspace of X and let A be a subset of Y. Denote by $\overline{A}^X$ the closure of A in the topological space X and by $\overline{A}^Y$ the closure of A in the topological space Y. Prove that $\overline{A}^Y \subset \overline{A}^X$. Show that in general $\overline{A}^Y \neq \overline{A}^X$.

7 PRODUCTS

Throughout this section let $(X_1, \mathfrak{I}_1), (X_2, \mathfrak{I}_2), \ldots, (X_n, \mathfrak{I}_n)$ be topological spaces and let $X = \prod_{i=1}^{n} X_i$. We wish to define a topology on X that may be regarded as the product of the topologies on the factors of X. Again our guide is the corresponding situation in metric spaces. If these topological spaces were metrizable, then there is a standard procedure for converting the product of the corresponding metric spaces into a metric space. In this resulting metric space, the open subsets of X are the unions of sets of the form $O_1 \times O_2 \times \ldots \times O_n$, where each O_i is an open subset of X_i. In the general situation, where $(X_i, \mathfrak{I}_i)$ may not be metrizable, one can show that the unions of the products of open sets will constitute a topology. This result is based on the following lemma.

Lemma 7.1 Let $\mathfrak{B}$ be a collection of subsets of a set X with the property that $\emptyset \in \mathfrak{B}$, $X \in \mathfrak{B}$, and a finite intersection of elements of $\mathfrak{B}$ is again in $\mathfrak{B}$. Then the collection $\mathfrak{I}$ of all subsets of X which are unions of elements of $\mathfrak{B}$ is a topology.

Proof. Clearly $\emptyset$ and X are in $\mathfrak{I}$. Suppose O and O' are in $\mathfrak{I}$. Then $O = \bigcup_{\alpha \in I} B_\alpha$, $O' = \bigcup_{\beta \in J} B_\beta$, where $B_\alpha \in \mathfrak{B}$ for $\alpha \in I$ and $B_\beta \in \mathfrak{B}$ for $\beta \in J$. Thus for $(\alpha, \beta) \in I \times J$, $B_\alpha \cap B_\beta \in \mathfrak{B}$ and hence

$$O \cap O' = \bigcup_{(\alpha,\beta) \in I \times J} (B_\alpha \cap B_\beta)$$

is in $\mathfrak{I}$. Finally a union of sets each of which is a union of sets of $\mathfrak{B}$ is again a union of sets of $\mathfrak{B}$ so that $\mathfrak{I}$ is a topology.

Since in the product set X the collection $\mathfrak{B}$ of subsets of X

that are unions of sets of the form $O_1 \times O_2 \times \ldots \times O_n$, each O_i an open subset of X_i, satisfies the conditions of this lemma we may state:

DEFINITION 7.2 The topological space $(X, \mathfrak{J})$, where $\mathfrak{J}$ is the collection of subsets of X that are unions of sets of the form

$$O_1 \times O_2 \times \ldots \times O_n,$$

each O_i an open subset of X_i, is called the *product* of the topological spaces $(X_i, \mathfrak{J}_i)$, $i = 1, 2, \ldots, n$.

In the future we shall often denote a topological space $(X, \mathfrak{J})$ simply by X. Thus, if we say, let $X_1, X_2, \ldots, X_n$ be topological spaces and $X = \prod_{i=1}^{n} X_i$, we shall mean that X is to be considered as the product of the topological spaces.

As was the case with metric spaces, the sets of the form $O_1 \times O_2 \times \cdots \times O_n$, O_i open in X_i, have been used as a "basis" for the open sets of X.

DEFINITION 7.3 Let X be a topological space and $\{O_\alpha\}_{\alpha \in I}$ a collection of open sets in X. $\{O_\alpha\}_{\alpha \in I}$ is called a *basis for the open sets of* X if each open set is a union of members of $\{O_\alpha\}_{\alpha \in I}$.

The next proposition characterizes the neighborhoods in the product space.

PROPOSITION 7.4 In a topological space $X = \prod_{i=1}^{n} X_i$, a subset N is a neighborhood of a point $a = (a_1, a_2, \ldots, a_n) \in N$ if and only if N contains a subset of the form $N_1 \times N_2 \times \ldots \times N_n$, where each N_i is a neighborhood of a_i.

Proof. First suppose that $N_1 \times N_2 \times \ldots \times N_n \subset N$ where each N_i is a neighborhood of a_i. By the definition of neighborhood in a topological space, each N_i contains an open set O_i containing a_i, hence N contains the open set $O_1 \times O_2 \times \ldots \times O_n$ containing a, and therefore N is a neighborhood of a. Conversely, suppose N is a neighborhood of a. Then N contains an open set O containing a. Since O is an open subset of the product space $X = \prod_{i=1}^{n} X_i$, we may write $O = \bigcup_{\alpha \in I} O_{\alpha,1} \times O_{\alpha,2} \times \ldots \times O_{\alpha,n}$, where for each i and each $\alpha \in I$, $O_{\alpha,i}$ is an open subset of X_i. Since $a \in O$, $a \in O_{\beta,1} \times O_{\beta,2} \times \ldots \times O_{\beta,n}$, for some $\beta \in I$, hence $a_i \in O_{\beta,i}$ for $i = 1, 2, \ldots, n$. But $O_{\beta,i}$ is open. Thus, if we set $N_i = O_{\beta,i}$, $i = 1, 2, \ldots, n$, N_i is a neighborhood of a_i and $N_1 \times N_2 \times \ldots \times N_n \subset O \subset N$.

DEFINITION 7.5 Let X be a topological space and $a \in X$. A collection $\mathfrak{N}_a$ of neighborhoods of a is called a *basis for the neighborhoods at* a if each neighborhood N of a contains a member of $\mathfrak{N}_a$.

Thus, if $a = (a_1, a_2, \ldots, a_n) \in X = \prod_{i=1}^{n} X_i$, a basis for the neighborhoods at a is the collection consisting of all subsets of the form $N_1 \times N_2 \times \ldots \times N_n$, where each N_i is a neighborhood of a_i.

Recall that in a product space the ith projection $p_i : X \to X_i$ is the function such that $p_i(a) = a_i$. If $O_i \in \mathfrak{I}_i$, then

$$p_i^{-1}(O_i) = X_1 \times \ldots \times X_{i-1} \times O_i \times X_{i+1} \times \ldots \times X_n.$$

Since this set is an open subset of X the projection maps are continuous.

A subset $O_1 \times O_2 \times \ldots \times O_n$ of X can be written as $p_1^{-1}(O_1) \cap \ldots \cap p_n^{-1}(O_n)$ so that we have a guide to the appropriate topology on an arbitrary product of topological spaces.

DEFINITION 7.6 Let $\{(X_\alpha, \mathfrak{I}_\alpha)\}_{\alpha \in A}$ be an indexed family of topological spaces. The topological product of this family is the set $X = \prod_{\alpha \in A} X_\alpha$ with the topology $\mathfrak{I}$ consisting of all unions of sets of the form $p_{\alpha_1}^{-1}(O_{\alpha_1}) \cap \ldots \cap$

$$p_{\alpha_k}^{-1}(O_{\alpha_k}), \text{ where } O_{\alpha_i} \in \mathfrak{I}_{\alpha_i}, i = 1, \ldots, k.$$

We have used as a basis for the topology $\mathfrak{I}$ the collection $\mathfrak{B}$ of sets of the form $p_{\alpha_1}^{-1}(O_{\alpha_1}) \cap \ldots \cap p_{\alpha_k}^{-1}(O_{\alpha_k})$, $O_{\alpha_i} \in \mathfrak{I}_{\alpha_i}$. That $\mathfrak{I}$ is a topology follows from the fact that $\emptyset \in \mathfrak{B}$, $X \in \mathfrak{B}$, and a finite intersection of elements of $\mathfrak{B}$ is again in $\mathfrak{B}$. Clearly this topology makes the projection maps continuous. Since any topology on X which makes the projection maps continuous must contain the sets of this form, the product topology is the weakest topology consistent with the continuity of the projection maps.

It is easily seen that, analogous to Proposition 7.4, a basis for the neighborhoods at a point x is the collection of sets of the form $p_{\alpha_1}^{-1}(N_{\alpha_1}) \cap \ldots \cap p_{\alpha_k}^{-1}(N_{\alpha_k})$, where N_{α_i} is a neighborhood of $p_{\alpha_i}(x) = x(\alpha_i) \in X_{\alpha_i}$ for $i = 1, \ldots, k$. In effect then, in the product space X we are saying that a point y is in a given neighborhood of x or is close to x if there is a finite set of indices $\{\alpha_1, \ldots, \alpha_k\}$ such that $y(\alpha_i)$ is close to $x(\alpha_i)$.

EXERCISES

1. Prove that a subset F of $X = \prod_{i=1}^{n} X_i$ is closed if and only if F is an intersection of sets, each of which is a finite union of sets of the form $F_1 \times F_2 \times \ldots \times F_n$, where each F_i is a closed subset of X_i. Formulate the corresponding statement in an arbitrary product of topological spaces.
2. Let $X = \Pi_{\alpha \in I} X_\alpha$ be the topological product of the family of spaces $\{X_\alpha\}_{\alpha \in I}$. Prove that a function $f: Y \to X$ from a space Y into the product X is continuous if and only if for each $\alpha \in I$ the function $f_\alpha = p_\alpha f: Y \to X_\alpha$ is continuous.
3. Let $\{X_\alpha\}_{\alpha \in I}$ be a family of spaces and let $I = J \cup K$, where J and K are disjoint and non-empty. Let $x \in \Pi_{\alpha \in J} X_\alpha$ be given. Define a function $\varphi_x: \Pi_{\alpha \in K} X_\alpha \to \Pi_{\alpha \in I} X_\alpha$ by setting for each $y \in \Pi_{\alpha \in K} X_\alpha$, $(\varphi_x(y))(\alpha) = y(\alpha)$ if $\alpha \in K$ and $(\varphi_x(y))(\alpha) = x(\alpha)$ if $\alpha \in J$. Prove that φ_x is continuous.
4. Let $\{X_\alpha\}_{\alpha \in I}$ and $\{Y_\alpha\}_{\alpha \in I}$ be two families of spaces indexed by the same indexing set I. For each $\alpha \in I$, let $f_\alpha: X_\alpha \to Y_\alpha$ be a continuous

function. Define $f:\Pi_{\alpha\in I} X_\alpha \to \Pi_{\alpha\in I} Y_\alpha$ by $(f(x))(\alpha) = f_\alpha(x(\alpha))$. Prove that f is continuous.

5. Let N be the set of positive integers. For each $n \in N$ let $X_n = \{0, 2\}$ with the discrete topology. Let $X = \Pi_{n\in N} X_n$. Define a function $f:X \to [0, 1]$ by setting $f(x) = \sum_{n=1}^{\infty} \frac{x(n)}{3^n}$. Prove that f is one-one and continuous. The image $f(X)$ is called the *Cantor set D* and consists of all real numbers $a \in [0, 1]$ which can be represented as triadic decimals $a = \sum_{n=1}^{\infty} \frac{a_n}{3^n}$ such that $a_n \in \{0, 2\}$ for all n. Given $a \in D$ define $(g(a))(n) = a_n$ so that $g(a) \in X$. Prove that g is a homeomorphism of D with X.
6. Prove that the family of open intervals with rational end points is a basis for the topology of the real line.

8 IDENTIFICATION TOPOLOGIES

Let R be the real line and S the unit circle defined by $S = \{(x, y) \mid (x, y) \in R^2, x^2 + y^2 = 1\}$. The function $p:R \to S$ defined by $p(t) = (\cos 2\pi t, \sin 2\pi t)$ maps R continuously onto S so that $p(t) = p(t')$, provided $t - t'$ is an integer. One may think of p as wrapping the real line around the circle so that the points which differ by an integer are identified or superimposed on each other. Furthermore, we shall see that the topology of S may be obtained from the topology of R in such a way as to make the mapping p an identification.

DEFINITION 8.1 Let $p:E \to B$ be a continuous function mapping the topological space E onto the topological space B. p is called an *identification* if for each subset U of B, $p^{-1}(U)$ open in E implies that U is open in B.

If $p:E \to B$ is an identification and $g:B \to Y$ is a continuous function defined on B, then g induces a continuous function $gp:E \to Y$. It turns out that frequently the reverse is true, that is, a continuous function $G:E \to Y$ will induce a continuous function $g:B \to Y$.

THEOREM 8.2 Let $p:E \to B$ be an identification and let $G:E \to Y$ be a continuous function such that for each $x, x' \in E$ with $p(x) = p(x')$, we also have $G(x) = G(x')$. Then for each $b \in B$ we may choose any $x \in p^{-1}(\{b\})$, define $g(b) = G(x)$, and the resulting function g is continuous.

Proof. First the definition of $g(b)$ does not depend on the choice of $x \in p^{-1}(\{b\})$, for if $x' \in p^{-1}(\{b\})$ then $p(x) = p(x')$ and $G(x) = G(x')$. Note that g is defined so that $gp = G$. Hence $G^{-1} = p^{-1}g^{-1}$. If O is an open subset of Y, then $G^{-1}(O)$ is open in E. But $G^{-1}(O) = p^{-1}(g^{-1}(O))$. Since p is an identification, $g^{-1}(O)$ is open in B and therefore g is continuous.

The hypothesis on the function G is that Gp^{-1} be well-defined or single-valued. The conclusion is then that the function g may be inserted in the diagram of Figure 9 and that commutativity will hold.

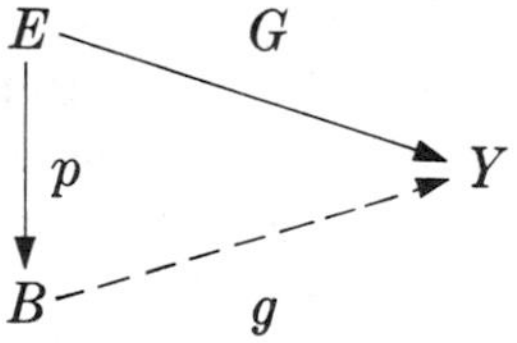

Figure 9

One may use an onto function $p:X \to Y$ from a topological space X to a set Y (without a topology) to construct a topology for Y so that p becomes an identification.

DEFINITION 8.3 Let $p:X \to Y$ be a function from a topological space X onto a set Y. The *identification topology on Y deter-*

> *mined by* p consists of those sets U such that $p^{-1}(U)$ is open in X.

Verification of the fact that this collection of sets is a topology depends on the behavior of p^{-1} with respect to unions and intersections. Once Y has been given the identification topology determined by p, p is an identification.

Let $f:X \to Y$ be a function from a set X to a set Y. Let $\sim_f$ be the relation defined on X by $x \sim_f x'$ if $f(x) = f(x')$. $\sim_f$ is an equivalence relation. Let $X/\sim_f$ be the collection of equivalence sets under this relation and let $\pi_f:X \to X/\sim_f$ be the function which maps each $x \in X$ into its equivalence class. π_f is an onto function. Now suppose that X is a topological space and give $X/\sim_f$ the identification topology determined by π_f. Let Y also be a topological space. Since $\pi_f(x) = \pi_f(x')$ if and only if $f(x) = f(x')$, f induces a continuous function $f^*:X/\sim_f \to Y$ such that $f = f^*\pi_f$. Furthermore f^* is one-one, for if $f^*(u) = f^*(u')$, with u, $u' \in X/\sim_f$, then for $x \in \pi_f^{-1}(\{u\})$, $x' \in \pi_f^{-1}\{u'\}$, $f(x) = f(x')$. Thus $x \sim_f x'$ or $u = \pi_f(x) = \pi_f(x') = u'$.

Consider the diagram

$$\begin{array}{ccc} & X & \\ {\scriptstyle \pi_f} \swarrow & & \searrow {\scriptstyle f} \\ X/\sim_f & \xrightarrow{f^*} & Y \end{array}$$

Let $\mathfrak{J}$ be the topology on $X/\sim_f$ and let $\mathcal{S}$ be the topology on Y. Since f^* is continuous, $f^{*-1}(\mathcal{S}) \subset \mathfrak{J}$, or equivalently, since f^* is one-one, $\mathcal{S} \subset f^*(\mathfrak{J})$. If $\mathcal{S}'$ were some other topology on Y so that f were continuous we would again have $\mathcal{S}' \subset f^*(\mathfrak{J})$. Thus the topology $\mathfrak{J}$ carried over to Y by f^* is the weakest or smallest topology such that f is continuous. Introducing the topologies into the diagram we obtain Figure 10, in which the inclusion map $i:(Y, f^*(\mathfrak{J})) \to (Y, \mathcal{S})$ is continuous.

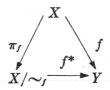

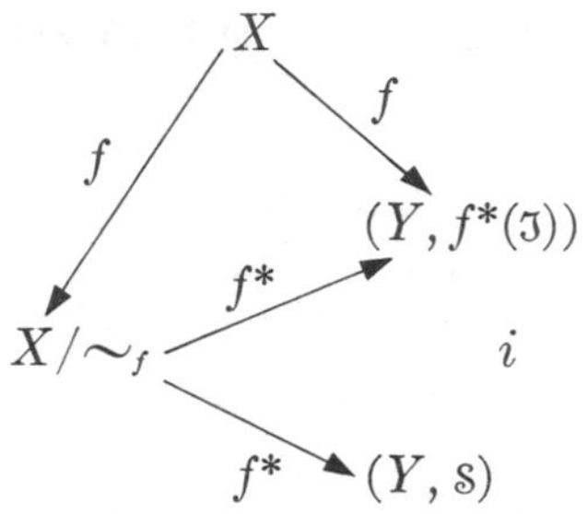

Figure 10

We shall conclude this section by considering some examples.

EXAMPLE 1 (The covering of the circle by the real line.) Let $p(t) = (\cos 2\pi t, \sin 2\pi t)$ so that $p: R \to S$ is a continuous mapping of the real line onto the circle. To show that p is an identification mapping, we must show that if $U \subset S$ is such that $p^{-1}(U)$ is open, then U is open. Let $x \in p^{-1}(U)$ and $s = p(x)$. x is the center of an open interval $O \subset p^{-1}(U)$ of length $2\varepsilon < 1$, which under p is mapped into an arc of S centered at s of length $4\pi\varepsilon$ and contained in U. This arc is an open ball in S with center s; hence U is open.

The function defined by $g(t) = (\cos 2\pi t, \sin 2\pi t, t)$ is a homeomorphism of the real line with a helix H in R^3. Let S be taken to be the set of points $(x, y, z) \in R^3$ defined by $x^2 + y^2 = 1$, $z = 0$. Then the projection of H onto S defined by $(\cos 2\pi t, \sin 2\pi t, t) \to (\cos 2\pi t, \sin 2\pi t, 0)$ is also an identification. This projection accounts for the literal sense in which the real line may be thought of as covering the circle.

Let f be any function defined and continuous on R. f is called *periodic* of period 1 if $f(t + 1) = f(t)$ for all $t \in R$. It follows that $f(t) = f(t')$, provided $t - t'$ is an integer, so that f induces a continuous function f^*, defined on the circle S such that $f^*(p(t)) = f(t)$.

EXAMPLE 2 (Shrinking a subset to a point.) Let X be a topological space and A a non-empty subset of X. Define a new untopologized set X/A as the union of $X - A$ and a new point a^*. Define a function $f:X \to X/A$ by $f(x) = x$ for $x \in X - A$, $f(x) = a^*$ for $x \in A$. Now give X/A the identification topology determined by f. This space is the space obtained by shrinking A to a point.

Let $\mathring{I} = \{0, 1\}$ be the boundary of the unit interval $I = [0, 1]$. Then $I/\mathring{I}$ is homeomorphic to a circle. In fact, by Theorem 8.2, the function $p(t) = (\cos 2\pi t, \sin 2\pi t)$, defined now for $t \in I$, must induce a continuous function $p^*: I/\mathring{I} \to S$. p^* is one-one and a basis for the open sets containing a^* is the totality of images of sets of the form $[0, \varepsilon) \cup (1 - \varepsilon, 1]$.

Shrinking the boundary of I to a point amounts to pasting the two end points together to make the single point a^* out of the boundary. If the boundary of a square is shrunk to a point, the resulting space turns out to be homeomorphic to the surface of a globe or a 2-sphere. One can even visualize this shrinking as a process in which an elastic sheet having a string in its boundary is deformed into a 2-sphere by gathering the string to a point.

Example 3 (Attaching a space X to a space Y.) Let X and Y be topological spaces and let A be a non-empty closed subset of X. Assume that X and Y are disjoint and that a continuous function $f:A \to Y$ is given. Form the set $(X - A) \cup Y$ and define a function $\varphi: X \cup Y \to (X - A) \cup Y$ by $\varphi(x) = f(x)$ for $x \in A$, $\varphi(x) = x$ for $x \in X - A$, and $\varphi(y) = y$ for $y \in Y$. Give $X \cup Y$ the topology in which a set is open (or closed) if and only if its intersections with both X and Y are open (or closed). φ is onto. Let $X \cup_f Y$ be the set $(X - A) \cup Y$ with the identification topology determined by φ.

If Y is a single point a^*, then attaching X to a^* by a function $f:A \to a^*$ is the same as shrinking A to a point. Let I^2 be the unit square in R^2 and let A be the union of its two vertical edges so that $A = \{(x, y) \mid (x, y) \in R^2$ and either $x = 0, 0 \leqq y \leqq 1$ or $x = 1, 0 \leqq y \leqq 1\}$. Let $Y = [0, 1]$ be the unit interval. Define $f: I^2 \to Y$ by $f(x, y) = y$. Then $I^2 \cup_f Y$ is a cylinder formed by identifying the two vertical edges of I^2.

EXERCISES

1. Let n be an integer. Let $\varphi_n: R \to R$ be the function from the real line into itself defined by $\varphi_n(x) = nx$. Let $p(t) = (\cos 2\pi t, \sin 2\pi t)$ as before. Show that φ_n induces a function $\Phi_n: S \to S$ of the circle into itself so that $\Phi_n p = p\varphi_n$. [Φ_n is said to wrap the circle around itself n times for positive n.]
2. A *torus* is the surface of a donut or an inflatable inner tube. It can be thought of as being generated by rotating a circle about a line in the plane of the circle, provided the circle and the plane do not intersect. Prove that if C is a circular cylinder with S_1 and S_2 as its boundary circles and S_1 and S_2 are identified by mapping them both homeomorphically onto some third circle S, giving a map $f: S_1 \cup S_2 \to S$, then $C \cup_f S$ is a torus.

    Define a relation in the plane R^2 by $(x, y) \sim (x', y')$ provided $x - x'$ and $y - y'$ are integers. Prove that $\sim$ is an equivalence relation. Let T be the collection of equivalence sets and $\varphi: R^2 \to T$ the mapping carrying each point into its equivalence set. Give T the identification topology determined by φ. Show that T is homeomorphic to a torus.
3. The unit disc is the set of points in R^2 given by $D = \{(x, y) \mid (x, y) \in R^2 \text{ and } x^2 + y^2 \leqq 1\}$. Its boundary in R^2 is the unit circle S. Let A be the subset of the circular cylinder $S \times [0, 1]$ given by $S \times \{1\}$. Prove that $S \times [0, 1]/A$ is homeomorphic to the disc D. ($S \times [0, 1]/A$ is the cone over S, see the next problem.)
4. Let X be a topological space and A the subset of $X \times [0, 1]$ given by $X \times \{1\}$. The space $X \times [0, 1]/A$ is called the *cone over* X. Denote this space by TX. Prove that if X and X' are homeomorphic, then so are TX and TX'.
5. Let X be a topological space and let p_0 and p_1 be two points not in $X \times [-1, 1]$. Let $f(x, -1) = p_0$ and $f(x, 1) = p_1$ for $x \in X$ define a mapping of $A = X \times (\{-1\} \cup \{1\}) \to \{p_0, p_1\}$. Let $Y = \{p_0, p_1\}$ have the discrete topology so that f is continuous. $X \times [-1, 1] \cup_f Y$ is called the *suspension* of X, and is denoted by SX. The *equator* is the image of $X \times \{0\}$ in SX. Prove that this subject of SX is homeomorphic to X. Prove that the image of $X \times [0, 1]$ in SX is homeomorphic to the cone over X, and that therefore the suspension of X

is two cones over X identified along the equator. Prove that the suspension of a circle is homeomorphic to the 2-sphere S^2.

9 CATEGORIES AND FUNCTORS

A great deal of the more recent work in topological spaces has involved the consideration of a collection of topological spaces and collections of continuous mappings between these spaces. It has proven to be extremely fruitful to formulate an abstract definition of the structure involved.

DEFINITION 9.1 A *category* C is a collection of objects A whose members are called the *objects* of the category and for each ordered pair (X, Y) of objects of the category a set $H(X, Y)$ called the *maps* of X into Y together with a rule of composition which associates to each $f \in H(X, Y)$ and $g \in H(Y, Z)$ a map $gf \in H(X, Z)$. This composition is associative, that is, if $f \in H(X, Y)$, $g \in H(Y, Z)$, $h \in H(Z, W)$, then $h(gf) = (hg)f$ and identities exist, that is, for each object $X \in A$ there is an element $1_X \in H(X, X)$ such that for all $g \in H(X, Y)$ $g1_X = g$ and for all $h \in H(W, X)$ $1_X h = h$.

In Chapter 1 we were concerned with the category C_S of sets and functions. That is, A_S is the class of all sets and for $X, Y \in A_S$, $H(X, Y)$ is the set of all functions from X to Y. For $X \in A_S$, 1_X is the identity mapping of X onto itself. In an obvious fashion one may obtain what we would call subcategories C' of C_S by taking as objects A' some specified collection of sets and for $X, Y \in A'$, $H'(X, Y)$ to be some specified set of functions from X to Y provided that we always include the identity mapping 1_X in $H(X, X)$ for each $X \in A'$ and for each ordered pair (X, Y) of A' include in $H'(X, Y)$ all functions f which can be written in the form hg for $h \in H'(W, Y)$, $g \in H'(X, W)$. For example A' might be all finite sets and $H'(X, Y)$ all functions from X to Y. In particular A' could contain a single set X and

$H'(X, X)$ could be all invertible functions.

In Chapter 2 the appropriate category was the category C_M of all metric spaces and continuous functions. Chapter 3 furnished us with our main example, namely the category C_T of all topological spaces and continuous mappings.

We shall include in some detail one more example of a category of algebraic objects.

DEFINITION 9.2 A *group* G is a set G together with a function which associates to each ordered pair g_1, g_2 of elements of G an element $g_1g_2 \in G$ such that:

(i) $g_1(g_2g_3) = (g_1g_2)g_3$ for $g_1, g_2, g_3 \in G$;

(ii) there is an element $e \in G$, called the *identity* such that for all $g \in G$, $eg = ge = g$;

(iii) for each $g \in G$ there is an element $g^{-1} \in G$ called the *inverse* of g, such that $gg^{-1} = g^{-1}g = e$.

A *homomorphism* f from a group G to a group K is a function $f:G \to K$ such that $f(e) = e'$ if e and e' are identities in G and K respectively and for all $g, g' \in G, f(gg') = f(g)f(g')$.

Let $\mathcal{G}$ be a collection of groups and for $G, K \in \mathcal{G}$ let $H(G, K)$ be the set of all homomorphisms of G into K. If we use the ordinary composition of functions to define for $f \in H(G, K)$ and $g \in H(K, L)$ an element $gf \in H(G, L)$, it is easily verified that we have constructed a category $C_{\mathcal{G}}$ of groups $\mathcal{G}$ and homomorphisms.

In Chapter 4 we shall associate to certain topological spaces a group called the fundamental group of the space.

A transformation from one category to another which preserves the structure of a category is called a "*functor*."

DEFINITION 9.3 Let C and C' be categories with objects A and A' respectively. A *functor** $F:C \to C'$ is a pair of func-

* Definition 9.3 defines a covariant functor. In further work one also needs to consider contravariant functors in which $F_2:H(X, Y) \to H'(F_1(Y), F_1(X))$ and $F_2(gf) = F_2(f)F_2(g)$.

tions F_1 and F_2 such that $F_1 : A \to A'$ and for each ordered pair X, Y of objects of A,

$$F_2 : H(X, Y) \to H'(F_1(X), F_1(Y))$$

so that $F_2(1_X) = 1_{F_1(X)}$ and $F_2(gf) = F_2(g)F_2(f)$ for $f \in H(X, Y)$, $g \in H(Y, Z)$.

In keeping with the notation of the examples let us denote an element $f \in H(X, Y)$ by $X \xrightarrow{f} Y$. If $F : C \to C'$ is a functor we have

$$F_1(X) \xrightarrow{F_2(f)} F_1(Y),$$

F_2 preserves identities, and if

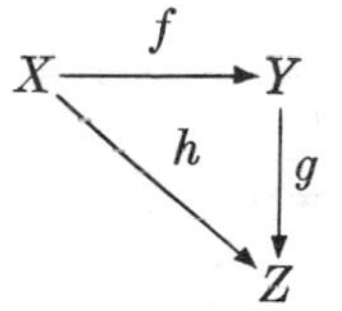

is commutative, then so is

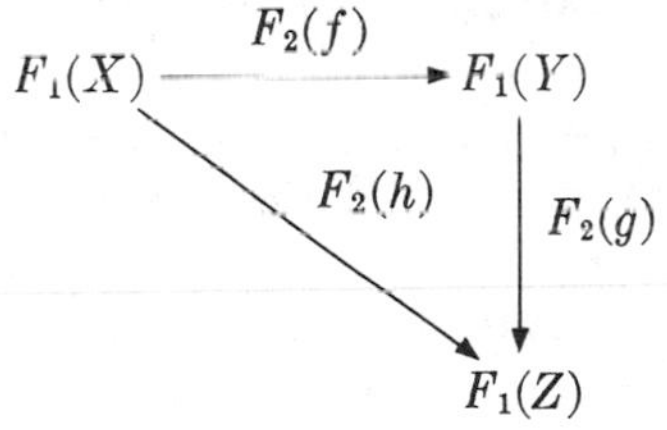

that is, F carries commutative diagrams into commutative diagrams.

The passage from a metric space (X, d) to its associated topological space $(X, \mathfrak{I})$ is an example of a functor from C_M to C_T. As another example of a functor, this time from C_T to itself, let Z be a fixed topological space. To each topological space $X \in C_T$ associate the topological space $F_1(X) = X \times Z$ and to each continuous function $f \in H(X, Y)$ associate the function

$F_2(f)$ defined by $(F_2(f))(x, z) = (f(x), z)$ for $(x, z) \in F_1(X)$. Then $F_2(f): F_1(X) \to F_1(Y)$ is continuous and it is easily verified that $F = (F_1, F_2)$ is a functor.

EXERCISES

1. Let $\{C_\alpha\}_{\alpha \in I}$ be an indexed family of categories with objects $\{A_\alpha\}_{\alpha \in I}$ and maps $\{H_\alpha(X, Y)\}_{\alpha \in I}$. Let $A = \Pi_{\alpha \in I} A_\alpha$. For $U, V \in A$ let $H(U, V) = \Pi_{\alpha \in I} (H_\alpha(U(\alpha), V(\alpha)))$. For $f \in H(U, V), g \in H(V, W)$ define gf by $gf(\alpha) = g(\alpha)f(\alpha)$. Prove that this yields a category $C = \Pi_{\alpha \in I} C_\alpha$ with objects A and maps $H(U, V)$.
2. Let C be a category with objects A. Let $e, f \in H(X, X)$ be such that $ge = g$, $gf = g$ for all $g \in H(X, Y)$ and $eh = h$, $fh = h$ for all $h \in H(W, X)$. Prove that $e = f$ and that therefore the identities are unique. Let $f \in H(X, Y)$ be such that there are maps $g, g' \in H(Y, X)$ with $gf = 1_X$ and $fg' = 1_Y$. Prove that $g = g'$ and that therefore f has a two-sided inverse $f^{-1} = g$. Such an f is called an *equivalence*. Prove that 1_X is an equivalence for all $X \in A$, if f is an equivalence so is f^{-1}, and if $f \in H(X, Y)$ and $f' \in H(Y, Z)$ are equivalences so is $f'f$. Verify that in the category C_T of topological spaces and continuous mappings the equivalences are the homeomorphisms. Prove that a functor carries equivalences into equivalences.
3. Let C_S be the category of sets and functions. Verify that the set of equivalences in $H(X, X)$ with the same rule of composition as in C_S is the group of one-one mappings of X onto itself. In general, verify that in any category C for each object X, the set of equivalences in $H(X, X)$ with the same rule of composition is a group.
4. Let A be a collection of pairs (X, Y) such that X is a topological space and Y is a subspace of X. Given (X, Y) and $(X', Y') \in A$ let $H((X, Y), (X', Y'))$ be the set of all continuous functions $f: X \to X'$ such that $f(Y) \subset Y'$. Construct a category with objects A and these maps. Verify that if for $(X, Y) \in A$ we set $F_1(X, Y) = Y$ and for $f \in H((X, Y), (X', Y'))$ we set $F_2(f) = f \mid Y$ then (F_1, F_2) is a functor.
5. Let C be a category whose objects are pairs (X, A) where X is a topological space and A is a non-empty closed subset of X, and whose maps $H((X, A), (Y, B))$ are continuous functions $f: X \to Y$

with $f(A) \subset B$. Define $F_1(X, A) = X/A$. Let $p_{(X, A)}: X \to X/A$ be the identification map. Prove that if $f \in H((X, A), (Y, B))$ then there is a continuous function $f^*: X/A \to Y/B$ such that the diagram

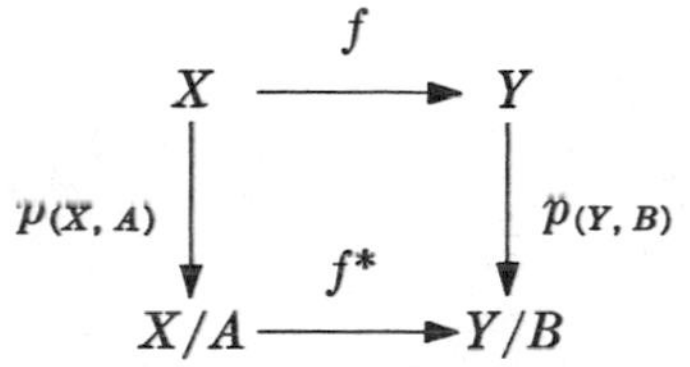

is commutative. Define $F_2(f) = f^*$ and prove that $F = (F_1, F_2)$ is a functor from the category C to the category C_T of topological spaces and continuous functions.

For further reading there are many excellent general texts including Kelley, *General Topology*, Dugundji, *Topology*, and Pervin, *Foundations of General Topology*.

CHAPTER 4

Connectedness

1 INTRODUCTION

A subspace of a topological space is "connected" if it is all "of one piece"; that is, if it is impossible to decompose the subspace into two disjoint non-empty open sets. The non-empty connected subsets of the real line are single points and intervals. The continuous image of a connected set is necessarily a connected set. A consequence of these two facts is the intermediate value theorem; that is, a continuous function $f\colon [a, b] \to R$ must assume all values between $f(a)$ and $f(b)$. A second type of connectedness is called "path-connectedness," by which it is meant that each pair of points may be "connected" by a "path" or "arc." Path-connectedness is a stronger condition than connectedness, since each path-connected topological space is connected, whereas the converse is false. A third type of connectedness that we shall consider is "simple connectedness." A topological space is simply connected if there are no holes in it to prevent the continuous shrinking of each closed arc to a point. The degree to which a given topological space fails to be simply connected may be measured by an algebraic topological invariant called the fundamental group of the space.

2 CONNECTEDNESS

DEFINITION 2.1 A topological space X is said to be *connected* if the only two subsets of X that are simultaneously open and closed are X itself and the empty set $\emptyset$. A topological space which is not connected is said to be *disconnected*.

Thus, a topological space X is disconnected if and only if there are two non-empty open subsets P and Q whose union is X and whose intersection is empty, for in this event P is the complement of Q and therefore both open and closed, whereas P is neither X nor $\emptyset$. Similarly, a topological space X is disconnected if and only if there are two non-empty closed subsets F and G whose union is X and whose intersection is empty.

Every subset A of a topological space X is itself a topological space in the relative topology. We say that the subset A is connected if the topological space A with the relative topology is connected, or what amounts to the same thing,

DEFINITION 2.2 A subset A of a topological space X is said to be *connected* if the only two subsets of A that are simultaneously relatively open and relatively closed in A are A and $\emptyset$.

Thus, the statement, A is connected, has the same meaning whether the reference is to A as a topological space or as a subspace of some larger topological space.

We shall shortly see that intervals such as $[a, b]$ and (a, b) are connected subsets of the real line R. As an example of a subset of the real line that is disconnected, let $A = [0, 1] \cup (2, 3)$.

$[0, 1]$ is a relatively closed subset of A since $[0, 1]$ is closed in R. At the same time $[0, 1]$ is a relatively open subset of A, since $[0, 1] = (-\frac{1}{2}, \frac{3}{2}) \cap A$. Finally, $[0, 1] \neq \emptyset$ and $[0, 1] \neq A$, hence A is disconnected. By the same token, the "open interval" $(2, 3)$ is also both relatively open and relatively closed in A.

It will be useful to have the following formulation of connectedness, or more precisely, disconnectedness.

LEMMA 2.3 Let A be a subspace of a topological space X. Then A is disconnected if and only if there exist two open subsets P and Q of X such that

$$A \subset P \cup Q,$$
$$P \cap Q \subset C(A),$$

and $P \cap A \neq \emptyset$, $Q \cap A \neq \emptyset$.

Proof. First, suppose that A is disconnected. Then there is a subset P' of A that is different from $\emptyset$ and from A and is both relatively open and relatively closed. This implies that the complement of P' in A, $C_A(P')$, is also different from $\emptyset$ and from A and relatively open. Thus $P' = P \cap A$ and $C_A(P') = Q \cap A$, where P and Q are open subsets of X. We therefore have that $A = P' \cup C_A(P') \subset P \cup Q$, for $P' \subset P$ and $C_A(P') \subset Q$, and also $P \cap Q \cap A = (P \cap A) \cap (Q \cap A) = P' \cap C_A(P') = \emptyset$ so that $P \cap Q \subset C(A)$. Finally, $P' = P \cap A$ and $C_A(P') = Q \cap A$ are non-empty.

Conversely, given open sets P and Q satisfying the stated conditions, set $P' = P \cap A$ and $Q' = Q \cap A$. Then $A = A \cap (P \cup Q) = (A \cap P) \cup (A \cap Q) = P' \cup Q'$ and $P' \cap Q' = (A \cap P) \cap (A \cap Q) = \emptyset$. Thus $P' = C_A(Q')$, and P' is both relatively open and relatively closed in A. Since $P' \neq \emptyset$ and $P' \neq A$ (for Q' is non-empty), A is disconnected.

A corresponding result also holds, using closed sets.

LEMMA 2.4 Let A be a subspace of a topological space X. Then A is disconnected if and only if there exist two closed subsets F and G of X such that

$$A \subset F \cup G,$$
$$F \cap G \subset C(A),$$
and $F \cap A \neq \emptyset$, $G \cap A \neq \emptyset$.

The next theorem asserts that connectedness is preserved under continuous mappings.

THEOREM 2.5 Let X and Y be topological spaces and let $f: X \to Y$ be continuous. If A is a connected subset of X, then $f(A)$ is a connected subset of Y.

Proof. Suppose $f(A)$ is not connected. Then there are open subsets P' and Q' of Y such that $f(A) \subset P' \cup Q'$, $P' \cap Q' \subset C(f(A))$, and $P' \cap f(A) \neq \emptyset$, $Q' \cap f(A) \neq \emptyset$. Since f is continuous, $P = f^{-1}(P')$ and $Q = f^{-1}(Q')$ are open subsets of X. But $A \subset f^{-1}(f(A)) \subset f^{-1}(P' \cup Q') = P \cup Q$. Also $P \cap Q = f^{-1}(P' \cap Q') \subset f^{-1}(Cf(A)) = C(f^{-1}(f(A))) \subset C(A)$. Finally, $P \cap A \neq \emptyset$, $Q \cap A \neq \emptyset$. Thus, A is not connected. It follows that if A is connected then $f(A)$ must also be connected.

COROLLARY 2.6 Let X and Y be topological spaces, let $f: X \to Y$ be a continuous mapping of X onto Y, and let X be connected; then Y is connected.

COROLLARY 2.7 Let X and Y be homeomorphic topological spaces, then X is connected if and only if Y is connected.

A property of a topological space is said to be a *topological property* if each topological space homeomorphic to the given space must also possess this property. Thus, Corollary 2.7 states that connectedness is a topological property.

Lemma 2.8 supplies an interesting characterization of connectedness, which will facilitate our proving that the product of two connected spaces is itself connected.

Lemma 2.8 Let $Y = \{0, 1\}$. A topological space X is connected if and only if the only continuous mappings $f:X \to Y$ are the constant mappings.

Proof. Let $f:X \to Y$ be a continuous non-constant mapping. Then $P = f^{-1}(\{0\})$ and $Q = f^{-1}(\{1\})$ are both non-empty. Thus, $P \neq \emptyset$ and $P \neq X$. $\{0\}$ and $\{1\}$ are open subsets of Y and f is continuous, therefore P and Q are open subsets of X. But $P = C(Q)$, so P is both open and closed and consequently X is disconnected. Thus, if X is connected, the only continuous mappings $f:X \to Y$ are constant mappings.

Conversely, suppose X is disconnected. Then there are non-empty open subsets P, Q of X such that $P \cap Q = \emptyset$ and $P \cup Q = X$. Define a mapping $f:X \to Y$ as follows: If $x \in P$, set $f(x) = 0$; if $x \in Q$, set $f(x) = 1$. f is continuous, for there are four open subsets, $\emptyset$, $\{0\}$, $\{1\}$, and Y of Y and $f^{-1}(\emptyset) = \emptyset$, $f^{-1}(\{0\}) = P$, $f^{-1}(\{1\}) = Q$, and $f^{-1}(Y) = X$, so that the inverse image of an open set is open.

Clearly, the role of the space $Y = \{0, 1\}$ in the above result could be played by any other topological space Z consisting of two points in which all subsets are open.

Theorem 2.9 Let X and Y be connected topological spaces. Then $X \times Y$ is connected.

Proof. We shall show that the only continuous mappings $f:X \times Y \to \{0, 1\}$ are constant mappings. Suppose, on the contrary, that there is a continuous mapping $f:X \times Y \to \{0, 1\}$ that is not constant. Then there are points $(x_0, y_0), (x_1, y_1) \in X \times Y$ such that $f(x_0, y_0) = 0$, $f(x_1, y_1) = 1$. If we picture $f(x, y)$ as a number attached to the point (x, y), then we have the situation depicted in Figure 11. Suppose $f(x_1, y_0) = 0$. We then define an "imbedding" $i_{x_1}:Y \to X \times Y$ by $i_{x_1}(y) = (x_1, y)$. i_{x_1} is continuous, hence the composite mapping $fi_{x_1}:Y \to \{0, 1\}$ is continuous (fi_{x_1} may be thought of as essentially f restricted to the points of the form (x_1, y); that is, the points lying above x_1 in

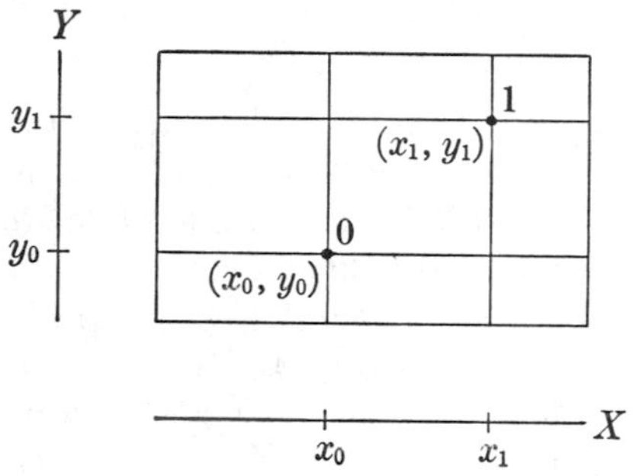

Figure 11

Figure 11.) But $(fi_{x_1})(y_0) = f(x_1, y_0) = 0$ and $(fi_{x_1})(y_1) = f(x_1, y_1) = 1$. Thus, in this case, there is a non-constant mapping of Y into $\{0, 1\}$, contradicting the connectedness of Y. Similarly, if $f(x_1, y_0) = 1$, we define an imbedding $i_{y_0}: X \to X \times Y$ by setting $i_{y_0}(x) = (x, y_0)$ and obtain a non-constant mapping $fi_{y_0}: X \to \{0, 1\}$, contradicting the connectedness of X. It follows that there are no non-constant mappings of $X \times Y$ into $\{0, 1\}$ and that therefore $X \times Y$ is connected.

Corollary 2.10 If $X_1, X_2, \ldots, X_n$ are connected topological spaces, then $\prod_{i=1}^{n} X_i$ is a connected topological space.

The main idea in the proof of Theorem 2.9 is that $f: X \times Y \to \{0, 1\}$ must remain constant on each of the connected subsets $\{x_0\} \times Y$ and $X \times \{y_1\}$. The same procedure allows us to show that in an arbitrary product $X = \Pi_{\alpha \in I} X_\alpha$ of connected spaces, altering a finite set of coordinates can not change the value of a continuous function $f: X \to \{0, 1\}$.

Lemma 2.11 Let $\{X_\alpha\}_{\alpha \in I}$ be an indexed family of topological spaces each of which is connected. Let x and x' be two points of $X = \Pi_{\alpha \in I} X_\alpha$ such that $x(\alpha) = x'(\alpha)$ except on a finite set of indices $I' \subset I$ and let $f: X \to \{0, 1\}$ be continuous. Then $f(x) = f(x')$.

Proof. We shall define an "imbedding" of $\Pi_{\alpha \in I'} X_\alpha$ into X. Let $J = I - I'$ so that $x(\alpha) = x'(\alpha)$ for $\alpha \in J$. Given $z \in \Pi_{\alpha \in I'} X_\alpha$, set $(j(z))(\alpha) = z(\alpha)$ for $\alpha \in I'$ and $(j(z))(\alpha) = x(\alpha)$ for $\alpha \in J$. Then $j: \Pi_{\alpha \in I'} X_\alpha \to X$ and j is continuous, for each of the functions $p_\alpha j$ is continuous (in fact, $p_\alpha j = p_\alpha$ for $\alpha \in I'$, $p_\alpha j$ is a constant function for $\alpha \in J$). Both x and x' are in the image of the connected set $\Pi_{\alpha \in I'} X_\alpha$ so that $f(x) = f(x')$.

Theorem 2.12 $X = \Pi_{\alpha \in I} X_\alpha$ is connected if each X_α is connected.

Proof. Again let $f: X \to \{0, 1\}$ be continuous and let $w, x \in X$ be such that $f(w) = 0$. We will show that $f(x) = 0$. $\{0\}$ is a neighborhood of 0, hence there is a neighborhood N of w such that $f(N) = 0$. It follows that there is a finite set of indices $I' = \{\alpha_1, \ldots, \alpha_k\}$ and neighborhoods N_{α_i} of $w(\alpha_i)$ in X_{α_i}, $i = 1, \ldots, k$, such that $p_{\alpha_1}^{-1}(N_{\alpha_1}) \cap \cdots \cap p_{\alpha_k}^{-1}(N_{\alpha_k}) \subset N$. Define a point $x' \in X$ by setting $x'(\alpha_i) = w(\alpha_i)$, $i = 1, \ldots, k$, $x'(\alpha) = x(\alpha)$ for all other $\alpha \in I$. Then $x' \in N$ so $f(x') = 0$. Since $x(\alpha) = x'(\alpha)$ except for $\alpha \in I'$, $f(x) = 0$.

EXERCISES

1. On the real line, prove that the set of non-zero numbers is not a connected set.
2. Let A and B be subsets of a topological space X. If A is connected, B is open and closed, and $A \cap B \neq \emptyset$, prove that $A \subset B$. [*Hint:* Assume $A \not\subset B$ and use the sets $P = A \cap B$ and $Q = A \cap C(B)$ to prove that A is not connected.]
3. Let A and B be connected subsets of a topological space X. If $A \cap B \neq \emptyset$, prove that $A \cup B$ is connected. [*Hint:* in the topological space $A \cup B$, show by using the result of Problem 2 that the only non-empty open and closed subset is $A \cup B$.]
4. Let A and B be non-empty subsets of a space X. Prove that $A \cup B$ is disconnected if $(\bar{A} \cap B) \cup (A \cap \bar{B}) = \emptyset$. Prove that X is con-

nected if and only if for every pair of non-empty subsets A and B of X such that $X = A \cup B$ we have $(\bar{A} \cap B) \cup (A \cap \bar{B}) \neq 0$.

5. Prove that a space X is connected if and only if for every non-empty subset A of X different from X we have Bdry $(A) \neq \emptyset$.

3 CONNECTEDNESS ON THE REAL LINE

In this section we shall define the term "interval" and prove that a non-empty subset of the real line is connected if and only if it is either a single point or an interval.

DEFINITION 3.1 A subset A of the real line is called an *interval* if A contains at least two distinct points, and if given points $a, b \in A$ with $a < b$, then for each point x such that $a < x < b$, it follows that $x \in A$.

Thus, an interval contains all points between any two of its points. It is a simple matter to verify that a closed interval $[a, b]$ or an open interval (a, b) is an interval in the sense of Definition 3.1. Other subsets of the real line that are intervals are defined in Definition 3.2.

DEFINITION 3.2 Let a be a real number. The subset of R consisting of all real numbers x such that $a < x$ is denoted by $(a, +\infty)$. The subset of R consisting of all real numbers x such that $a \leqq x$ is denoted by $[a, +\infty)$. The subset of R consisting of all real numbers x such that $x < a$ is denoted by $(-\infty, a)$. The subset of R consisting of all real numbers x such that $x \leqq a$ is denoted by $(-\infty, a]$.

Let b be a second real number with $a < b$. The subset of R consisting of all real numbers x such that $a < x \leqq b$ is denoted by $(a, b]$. The subset of R consisting of all real numbers x such that $a \leqq x < b$ is denoted by $[a, b)$.

We shall also denote R itself by $(-\infty, +\infty)$.

The subsets of R that have been mentioned in this section exhaust the collection of intervals.

THEOREM 3.3 A subset A of the real numbers is an interval if and only if it is of one of the following forms: (a, b); $[a, b)$; $(a, b]$; $[a, b]$; $(-\infty, a)$; $(-\infty, a]$; $(a, +\infty)$; $[a, +\infty)$; $(-\infty, +\infty)$.

Proof. We leave it to the reader to verify that each of these nine types of sets is an interval and shall prove the "only if" part of the theorem. Suppose A is an interval. We first note that if a point $x \notin A$, then either x is a lower bound of A or an upper bound of A, for otherwise there would be points $a, b \in A$ with $a < x < b$ and we would obtain the contradiction $x \in A$. We shall, consequently, distinguish four cases.

Case 1. A has neither an upper bound nor a lower bound. In this case $C(A)$ must be empty so that $A = (-\infty, +\infty)$.

Case 2. A has an upper bound but no lower bound. Since an interval is non-empty, A has a least upper bound a. We claim that if $x < a$, then $x \in A$. For, suppose $x < a$, then there is a point $a' \in A$ with $x < a' \leqq a$ (for otherwise a would not be a least upper bound). Since x cannot be a lower bound of A there is a point $b \in A$ with $b < x$. But $b < x < a'$ and $a', b \in A$ imply that $x \in A$. We have thus shown that $(-\infty, a) \subset A$. On the other hand, for $x > a$, $x \notin A$. It follows that A is either of the form $(-\infty, a]$ or $(-\infty, a)$, depending on whether $a \in A$ or $a \notin A$.

Case 3. A has a lower bound but no upper bound. By reasoning similar to that of Case 2, one shows that A is either of the form $[a, +\infty)$ or $(a, +\infty)$, where a is the greatest lower bound of A.

Case 4. A has a lower bound and an upper bound. Let a be the greatest lower bound of A and let b be the least upper bound of A. Since A contains at least two distinct points, $a < b$. A point x, if it is to lie in A, must therefore lie in $[a, b]$, so that $A \subset [a, b]$. We claim that

$a < x < b$ implies that $x \in A$. This implication follows from the fact that for any such point x, there must be points a' and b' with $a', b' \in A$ and $a \leqq a' < x < b' \leqq b$. Hence $(a, b) \subset A \subset [a, b]$. Consequently, A must be of one of the four forms (a, b), $[a, b)$, $(a, b]$, or $[a, b]$, depending on which, if any, of the two points a, b belong to A.

We shall now prove that apart from the empty set and single points, the only connected subsets of the real line are intervals.

THEOREM 3.4 A subset A of the real line that contains at least two distinct points is connected if and only if it is an interval.

Proof. We shall first show that if A is not an interval then it is not connected. If A is not an interval, then there are points a, b, c with $a < c < b$ and $a, b \in A$, whereas $c \notin A$. Let $P = (-\infty, c)$, $Q = (c, +\infty)$. P and Q are open subsets of the real line that satisfy the conditions of Lemma 2.3; hence A is not connected.

Conversely, we shall show that if A is not connected then A is not an interval. If A is not connected, by Lemma 2.4, there are closed subsets F and G of the real line such that $A \subset F \cup G$, $F \cap G \subset C(A)$ and both F and G contain a point of A. Assume that the notation is such that there is a point $a \in A \cap F$ and a point $b \in A \cap G$ with $a < b$. We shall find a point between a and b that is not in A. Let $G' = G \cap [a, b]$. Then G' is a closed non-empty subset of the real line and, consequently, contains its greatest lower bound c. We cannot have $a = c$, for then $A \cap F \cap G \neq \emptyset$, contradicting $F \cap G \subset C(A)$. Thus, $a < c$. Next, let $F' = F \cap [a, c]$. F' is also a closed non-empty subset of the real line and therefore contains its least upper bound d. In the event that $c = d$ we have $c \in F \cap G$, hence $c \notin A$ and A is not an interval. Otherwise $d < c$ and $(d, c) \cap (F \cup G) = \emptyset$, so that $(d, c) \cap A = \emptyset$, and again A does not contain a point between a and b and is therefore not an interval.

EXERCISES

1. Let $f: R \to R$ be continuous. Prove that the image under f of each interval is either a single point or an interval.
2. Prove that a homeomorphism $f: [a, b] \to [a, b]$ carries end points into end points.
3. Let A and B be subsets of R. A function $f: A \to B$ is called *monotone increasing* if $x, y \in A$ and $x < y$ imply $f(x) < f(y)$.
   (a) Let $f: A \to B$ be monotone increasing. Prove that $f: A \to B$ is one-one.
   (b) Let $f: [a, b] \to [f(a), f(b)]$ be monotone increasing and continuous. Prove that f is a homeomorphism.

4 SOME APPLICATIONS OF CONNECTEDNESS

THEOREM 4.1 (Intermediate-Value Theorem). Let $f: [a, b] \to R$ be continuous and let $f(a) \neq f(b)$. Then for each number V between $f(a)$ and $f(b)$ there is a point $v \in [a, b]$ such that $f(v) = V$.

Proof. $[a, b]$ is connected, hence $f([a, b])$ is connected and is therefore an interval. Now, $f(a), f(b) \in f([a, b])$. Thus if V is between $f(a)$ and $f(b)$, since $f([a, b])$ is an interval, $V \in f([a, b])$; that is, there is a $v \in [a, b]$ such that $f(v) = V$.

Theorem 4.1 states that for each V between $f(a)$ and $f(b)$, the horizontal line $y = V$ intersects the graph of $y = f(x)$ at some point (v, V) with $a < v < b$, as indicated in Figure 12.

If the domain of a continuous real-valued function contains an interval $[a, b]$, then its restriction to $[a, b]$ is continuous and we can assert that f must assume at least once each value between $f(a)$ and $f(b)$ over the interval $[a, b]$.

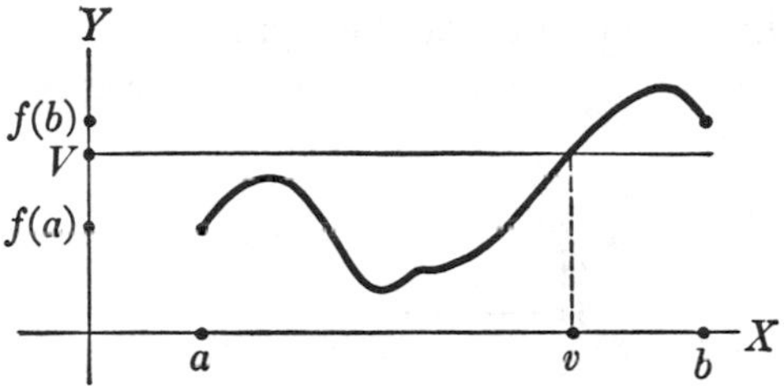

Figure 12

As a special case of the intermediate-value theorem, namely $V = 0$, we have

COROLLARY 4.2 Let $f: [a, b] \to R$ be continuous. If $f(a)f(b) < 0$, then there is an $x \in [a, b]$ such that $f(x) = 0$.

COROLLARY 4.3 (Fixed-Point Theorem). Let $f: [0, 1] \to [0, 1]$ be continuous. Then there is a $z \in [0, 1]$ such that $f(z) = z$.

Proof. In the event that $f(0) = 0$ or $f(1) = 1$, the theorem is certainly true. Thus, it suffices to consider the case in which $f(0) > 0$ and $f(1) < 1$. Let $g: [0, 1] \to R$ be defined by

$$g(x) = x - f(x),$$

(therefore, if $g(z) = 0$, $f(z) = z$). g is continuous and $g(0) = -f(0) < 0$, whereas $g(1) = 1 - f(1) > 0$. Consequently, by Corollary 4.2, there is a $z \in [0, 1]$ such that $g(z) = 0$, whence $f(z) = z$.

We may interpret this theorem geometrically. Since $f: [0, 1] \to [0, 1]$, the graph of $y = f(x)$ is contained in the unit square defined by $0 \leqq x \leqq 1$, $0 \leqq y \leqq 1$. The point $(z, f(z))$ given by the theorem lies on both the graph of $y = f(x)$ and the line $y = x$. Hence the theorem asserts that the graph of $y = f(x)$ intersects the line $y = x$ in this square (see Figure 13), or equivalently, that in order for the curve which constitutes the graph to connect a point on the left-hand edge of the square with a point on the right-hand edge of the square, the curve must intersect the diagonal of the square pictured in Figure 13.

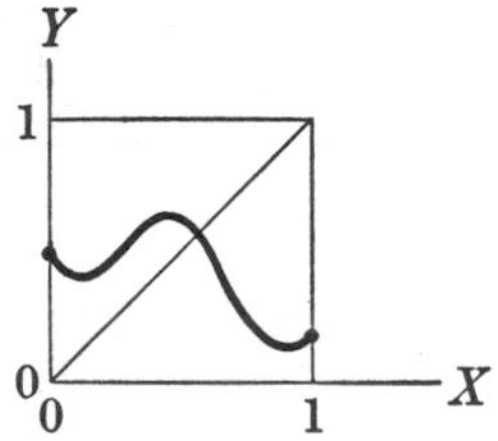

Figure 13

The reason for calling Theorem 4.3 a fixed-point theorem is that, if we think of $f:[0, 1] \to [0, 1]$ as a transformation that carries each point x of $[0, 1]$ into the point $f(x)$ of $[0, 1]$, then to say that $f(z) = z$ is to say that the transformation f leaves z "fixed."

There are many so-called "fixed-point" theorems, of which Corollary 4.3 is undoubtedly the simplest. In general, a fixed-point theorem is one that states that for a specified topological space X each continuous function $f:X \to X$ possesses a fixed point; that is, there is necessarily a $z \in X$ such that $f(z) = z$. One of the convenient facts about a fixed-point theorem is that if X and Y are homeomorphic topological spaces and a fixed-point theorem is true for X, then it is also true for Y.

THEOREM 4.4 Let X and Y be homeomorphic topological spaces. Then each continuous function $h:X \to X$ possesses a fixed point if and only if each continuous function $k:Y \to Y$ possesses a fixed point.

Proof. Let $f:X \to Y$ and $g:Y \to X$ be a pair of continuous inverse functions. Let $k:Y \to Y$ be a continuous function so that we have the diagram

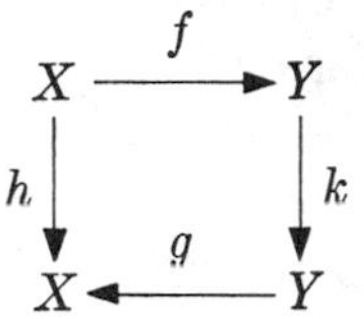

and suppose that each continuous function $h: X \to X$ possesses a fixed point. Then the function $h = gkf: X \to X$ is continuous and there is a $z \subset X$ such that $h(z) = z$. Let $w = f(z)$. We have

$$k(w) = k(f(z)) = fg(k(f(z))) = f(h(z)) = f(z) = w.$$

Thus, w is a fixed point of k. Since the hypotheses are symmetric with regard to X and Y, it also follows that if each continuous function $k: Y \to Y$ has a fixed point then so does each continuous function $h: X \to X$.

Any two closed intervals $[a, b]$ and $[c, d]$ are homeomorphic. Since a fixed-point theorem holds for $[0, 1]$, we obtain

COROLLARY 4.5 Let $f: [a, b] \to [a, b]$ be continuous. Then there is a $z \in [a, b]$ such that $f(z) = z$.

Theorem 4.3 is a special case of the "Brouwer Fixed-Point Theorem," which we shall now state. Recall that in R^n, the unit n-cube I^n is defined as the set of points $(x_1, x_2, \ldots, x_n)$ whose coordinates satisfy the inequalities $0 \leqq x_i \leqq 1$, for $i = 1, 2, \ldots, n$.

THEOREM 4.6 (Brouwer Fixed-Point Theorem). Let $f: I^n \to I^n$ be continuous. Then there is a point $z \in I^n$ such that $f(z) = z$.

We shall not prove this theorem. However, one can supply a very suggestive argument for the truth of the theorem in the case $n = 2$. To this end we may, on the basis of Theorem 4.5, work with a topological space homeomorphic to I^2. If we think of I^2 as being a surface constructed of elastic material, we may conceive of a deformation or stretching by which we obtain a surface that is a disc; that is, the set of points (x_1, x_2) in the plane whose coordinates satisfy the inequality $x_1^2 + x_2^2 \leqq 1$. Thus, the disc is homeomorphic with I^2, and we may argue the validity of the fixed-point theorem with regard to the disc.

Let g be a continuous transformation of this disc into itself. Suppose that it were possible that for each point x of the disc, we had $g(x) \neq x$. Then for each point x in the disc, there would be a unique half-line L_x emanating from $g(x)$ and passing through x (see Figure 14). The half-line L_x will contain a point

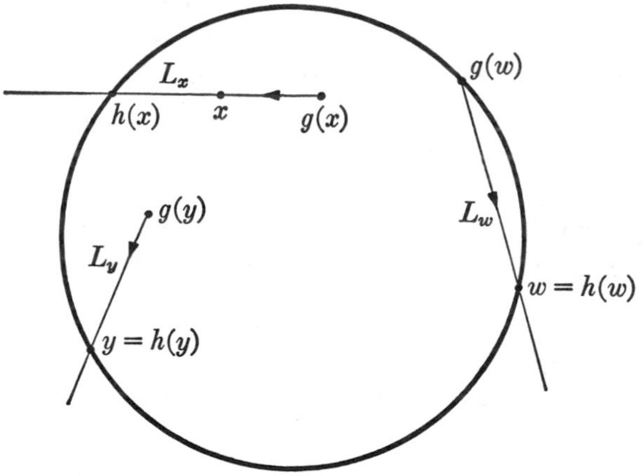

Figure 14

on the boundary of the disc other than $g(x)$. Let us call this point $h(x)$. In particular, if y is a boundary point of the disc, then $h(y) = y$. This is true even if $g(y)$ itself is a boundary point, as may be seen by examining the various cases depicted in Figure 14. Using the given transformation g we have thus constructed a new transformation h, which has the property that it carries each point of the disc into a boundary point and leaves each boundary point fixed (h is called a "retraction" since it retracts or pulls the interior of the disc onto its boundary while leaving the boundary fixed).

We next argue that the transformation h is continuous, for the image $h(x)$ will vary by a small amount if we suitably restrict the variation of x. Though it is by no means simple to prove that no continuous transformation such as h can exist, our intuition should tell us that none can. For if there were a function such as h we should be able to retract the head of a drum onto the rim, although intuitively we know that we can do so only by ripping the drum head someplace; that is, by introducing a discontinuity. Since there is no function such as the retraction h, we have obtained a contradiction, and therefore our supposition that g did not have a fixed point is untenable.

Another application of the intermediate-value theorem relates to the concept of antipodal points on spheres. Let us recall that the n-sphere, S^n, is the set of points $(x_1, x_2, \ldots, x_{n+1})$ in R^{n+1} satisfying the equation $x_1^2 + x_2^2 + \ldots x_{n+1}^2 = 1$, the topology of S^n being the relative topology. If $(x_1, x_2, \ldots, x_{n+1})$ is in S^n, the pair of points $(x_1, x_2, \ldots, x_{n+1})$ and $(-x_1, -x_2, \ldots, -x_{n+1})$ are called a *pair of antipodal points*. Given $x = (x_1, x_2, \ldots, x_{n+1}) \in S^n$, it is convenient to denote $(-x_1, -x_2, \ldots, -x_{n+1})$ by $-x$ and call $-x$ the *antipodal point* of x. A pair x, $-x$ of antipodal points is the pair of end points of a diameter of the sphere. We shall be particularly interested in the 1-sphere, S^1, which is a circle.

Consider a continuous function $f:S^1 \to R$. If we define $F:S^1 \to R$ by $F(x) = f(x) - f(-x)$ for $x \in S^1$, we can show

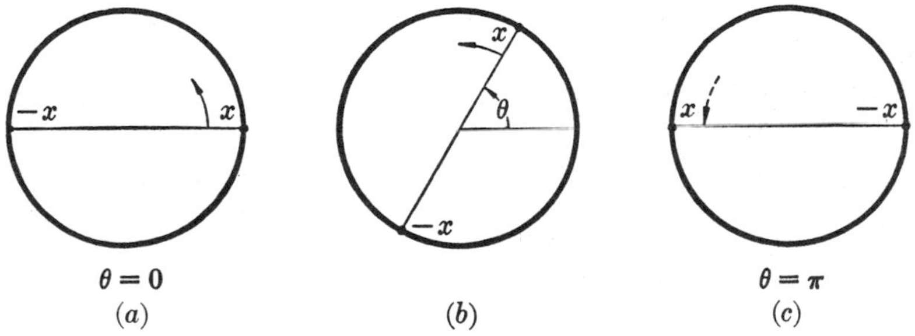

Figure 15

that $F(z) = 0$ for some $z \in S^1$; that is, $f(z) = f(-z)$, or f has the same value at one or more pairs of antipodal points. The proof of this fact is motivated by the consideration that a value of F is determined by a diameter of the circle and a designation of one of its extremities as x and the other as $-x$. If we rotate this diameter through π radians, as indicated in Figure 15, then the initial value of F corresponding to Figure 15a is opposite in sign to the final value of F corresponding to Figure 15c. But F is continuous, so its value must be zero for some position of the diameter corresponding to a value of θ with $0 \leqq \theta \leqq \pi$, where θ is the angle through which the diameter has been rotated. Thus $F(z) = 0$ and

THEOREM 4.7 Let $f: S^1 \to R$ be continuous, then there exists a pair of antipodal points $z, -z \in S^1$ such that $f(z) = f(-z)$.

Theorem 4.7 is the case $n = 1$ of the Borsuk-Ulam Theorem.

THEOREM 4.8 Let $f: S^n \to R^n$ be continuous. Then there exists a pair of antipodal points $z, -z \in S^n$ such that $f(z) = f(-z)$.

We shall not prove this theorem. The case $n = 2$ answers a question about map making. The 2-sphere S^2 may be thought of as the surface of a globe. In this case the Borsuk-Ulam Theorem gives a negative answer to the question, "Is it possible to draw a map of the surface of the earth on a flat sheet of paper so that distinct points on the surface of the earth correspond to distinct points on the map, and nearby points on the surface of the earth correspond to nearby points on the map?" The reason the answer to this question is "no" is that otherwise the existence of such a correspondence would imply the existence of a continuous function $f: S^2 \to R^2$ that was one-one, and this possibility is ruled out by the Borsuk-Ulam Theorem.

EXERCISES

1. Prove that a polynomial of odd degree considered as a function from the reals to the reals has at least one real root.

2. Let $f:[a, b] \to R$ be a continuous function from a closed interval into the reals. Let $U = f(u)$ and $V = f(v)$ be such that $U \leqq f(x) \leqq V$ for all $x \in [a, b]$. Prove that there is a w between u and v such that $f(w) \cdot (b - a) = \int_a^b f(t)\, dt$.

3. Let $F:R^2 \to R$ be a real-valued function defined and continuous on the plane. For each continuous function $f:[a, b] \to R$ we may define a new continuous function $Kf:[a, b] \to R$ by setting $Kf(t) = \int_a^t F(x, f(x))\, dx$, $t \in [a, b]$. Thus, if S is the set of continuous real-valued functions defined on $[a, b]$, K defines a transformation of S into itself. Prove that an element $g \in S$ is a fixed point of K if and only if g satisfies the differential equation $g'(x) = F(x, g(x))$ with initial condition $g(a) = 0$.

4. Prove that the mapping $p:R \to S^1$ defined by $p(t) = (\cos t, \sin t)$ for $t \in R$ is continuous and that therefore, for each continuous function $g:S^1 \to R$, there is a continuous function $\varphi:R \to R$ such that the diagram

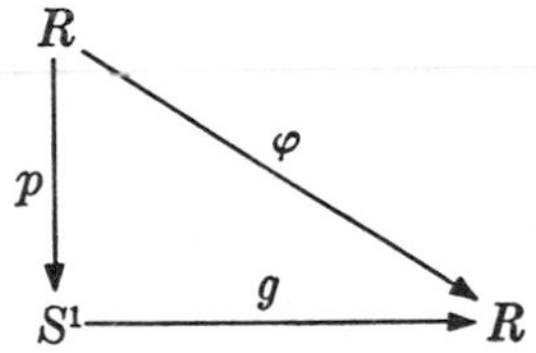

is commutative. Let $f:S^1 \to R$ be continuous and define $F:S^1 \to R$ by $F(x) = f(x) - f(-x)$ for $x \in S^1$. Prove that $(Fp)(t) = -(Fp)(t + \pi)$, and that therefore there is a $z \in [0, \pi]$ such that $(Fp)(z) = 0$. Then show that if $x = p(z)$, $f(x) = f(-x)$, thereby proving Theorem 4.7.

5 COMPONENTS AND LOCAL CONNECTEDNESS

In any topological space X, each point $a \in X$ belongs to a maximal connected subset of X called the "component of a."

THEOREM 5.1 Let X be a topological space. For each point $a \in X$ there is a non-empty subset $\text{Cmp}(a)$, called the *component* of a, with the property that $\text{Cmp}(a)$ is connected and if D is any connected subset of X containing a, then $D \subset \text{Cmp}(a)$.

Proof. There are connected subsets of X containing a for $\{a\}$ is such a subset. Let I be an indexing set for the family of connected subsets $\{D_\alpha\}_{\alpha \in I}$ containing a. We set $\text{Cmp}(a) = \bigcup_{\alpha \in I} D_\alpha$. Thus, if D is any connected subset of X containing a, $D = D_\beta$ for some $\beta \in I$, whence $D \subset \text{Cmp}(a)$. It remains to prove that $\text{Cmp}(a)$ is connected. Suppose it is not. Then there are non-empty relatively open subsets A and B of $\text{Cmp}(a)$ such that $A \cap B = \emptyset$ and $A \cup B = \text{Cmp}(a)$. Assume the notation is such that $a \in A$ and let b be a point of B. Since $b \in \text{Cmp}(a)$, $b \in D_\gamma$ for some connected subset D_γ of X containing a. Now $D_\gamma \subset \text{Cmp}(a)$, hence $A' = A \cap D_\gamma$ and $B' = B \cap D_\gamma$ are non-empty relatively open subsets of D_γ. Furthermore, $A' \cap B' \subset A \cap B = \emptyset$ and $A' \cup B' = D_\gamma \cap (A \cup B) = D_\gamma$. Consequently, the supposition that $\text{Cmp}(a)$ is not connected yields the contradiction that D_γ is not connected. Therefore $\text{Cmp}(a)$ is connected.

LEMMA 5.2 In a topological space X, let $b \in \text{Cmp}(a)$. Then $\text{Cmp}(b) = \text{Cmp}(a)$.

Proof. Since $b \in \text{Cmp}(a)$ and $\text{Cmp}(a)$ is a connected set containing b, by Theorem 5.1, $\text{Cmp}(a) \subset \text{Cmp}(b)$. But $a \in \text{Cmp}(a)$, hence $a \in \text{Cmp}(b)$, so, by the same reasoning it follows that $\text{Cmp}(b) \subset \text{Cmp}(a)$ and therefore $\text{Cmp}(a) = \text{Cmp}(b)$.

COROLLARY 5.3 In a topological space X, define $a \sim b$ if $b \in \text{Cmp}(a)$. Then $\sim$ is an equivalence relation.

Proof. Since $\{a\}$ is connected, $a \in \text{Cmp}(a)$ so $a \sim a$. If $a \sim b$ or $b \in \text{Cmp}(a)$, then by 5.2, $\text{Cmp}(a) = \text{Cmp}(b)$. We have already seen that $a \in \text{Cmp}(a)$, so $a \in \text{Cmp}(b)$ and $b \sim a$. Finally, if $a \sim b$ and $b \sim c$, then as before, $\text{Cmp}(a) = \text{Cmp}(b) = \text{Cmp}(c)$, whence $c \in \text{Cmp}(a)$ and $a \sim c$.

A subset of X that is a component of some point $a \in X$ is called a *component* of X. The components are the equivalence sets under the relation $b \in \text{Cmp}(a)$. They constitute a partition of X into maximal connected subsets in the sense of the following definition.

DEFINITION 5.4 Let X be a set and $\{P_\alpha\}_{\alpha \in I}$ an indexed family of non-empty subsets of X. $\{P_\alpha\}_{\alpha \in I}$ is called a *partition* of X if:

(i) $X = \bigcup_{\alpha \in I} P_\alpha$;

(ii) If $\alpha, \beta \in I$, $\alpha \neq \beta$, then $P_\alpha \cap P_\beta = \emptyset$.

THEOREM 5.5 Let A be a connected subset of a topological space X and let $A \subset B \subset \bar{A}$. Then B is also connected.

Proof. We shall show that if B is not connected then A is not connected. For suppose there are open subsets P, Q of X such that $P \cap Q \subset C(B)$, $B \subset P \cup Q$, $P \cap B \neq \emptyset$, and $Q \cap B \neq \emptyset$. It would follow that $A \subset P \cup Q$ and since $C(B) \subset C(A)$, $P \cap Q \subset C(A)$. To prove that A is not connected we must show that $P \cap A \neq \emptyset$ and $Q \cap A \neq \emptyset$. If $P \cap A = \emptyset$, then A would be contained in the closed set $C(P)$, hence $\bar{A} \subset C(P)$ or $P \cap \bar{A} = \emptyset$. But this last relation would imply that $P \cap B = \emptyset$. Thus, $P \cap A \neq \emptyset$. Similarly, $Q \cap A \neq \emptyset$.

COROLLARY 5.6 The closure of a connected set is connected.

COROLLARY 5.7 In a topological space, each component is a closed set.

Proof. Let A be a component, say $A = \text{Cmp}(a)$. Then $\overline{A}$ is a connected set containing a and therefore $\overline{A} \subset \text{Cmp}(a) = A$. But $A \subset \overline{A}$, hence in this case $A = \overline{A}$ and A is closed.

It might be thought that a component must also be an open set, but it need not be as the following example will indicate. Let X be the subspace of the real line consisting of the points 0 and all numbers of the form $\frac{1}{n}$, n a positive integer. The only connected set containing 0 is $\{0\}$, thus $\text{Cmp}(0) = \{0\}$. On the other hand $\{0\}$ is not a neighborhood of 0 in X and hence $\{0\}$ is not an open subset of X.

A sufficient condition for the components in a space to be open is that the space be "locally connected."

DEFINITION 5.8 A topological space X is said to be *locally connected at a point* $a \in X$ if each neighborhood N of a contains a connected neighborhood U of a. A topological space X is said to be *locally connected* if it is locally connected at each of its points.

LEMMA 5.9 Let X be a locally connected topological space and let Q be a component. Then Q is an open set.

Proof. Let $a \in Q$. Since X is locally connected there is at least one connected neighborhood N of a. But $Q = \text{Cmp}(a)$, hence by Theorem 5.1, $N \subset Q$, which, in turn, implies that Q is a neighborhood of a. Thus, Q is a neighborhood of each of its points and therefore Q is open.

If X is locally connected at a then there are "arbitrarily small" connected neighborhoods of a, for, given any neighborhood N of a, there is a connected neighborhood $U \subset N$ that is at least as "small" as N. An equivalent formulation of local connectedness is obtained by utilizing the concept of basis for the neighborhoods at a.

LEMMA 5.10 A topological space is locally connected at a point $a \in X$ if and only if there is a basis for the neighborhoods at a composed of connected subsets of X.

Proof. First, suppose that X is locally connected at a and let U_a be the collection of connected neighborhoods of a. Since every neighborhood N of a contains an element of U_a, U_a is a basis for the neighborhoods at a. Conversely, if there is a basis U_a for the neighborhoods of a consisting of connected sets, each neighborhood N must contain an element of U_a and therefore X is locally connected at a.

EXERCISES

1. Prove that a non-empty connected subset of a topological space that is both open and closed is a component.
2. Let X be a topological space that has a finite number of components. Prove that each component of X is both open and closed.
3. Verify that local connectedness is a topological property, but the continuous image of a locally connected space need not be locally connected.
4. Let X and Y be homeomorphic topological spaces. Prove that any homeomorphism $f: X \to Y$ establishes a one-one correspondence between the components of X and the components of Y.
5. Prove that the product of two locally connected topological spaces is locally connected.
6. Prove that Euclidean n-space R^n and the standard n-cube I^n are locally connected.

6 PATH-CONNECTED TOPOLOGICAL SPACES

In the three-dimensional geometry of the calculus, one often discusses a curve in terms of a parametric representation, usually written $x = f(t)$, $y = g(t)$, $z = h(t)$. If not stated explicitly, it is generally understood that the three functions f, g, h are at least continuous, if not differentiable over some common interval $[a, b]$ as their domain, and therefore $F(t) = (f(t), g(t), h(t))$ defines a continuous function $F:[a, b] \to R^3$. The curve in question is, from this viewpoint, the image of $[a, b]$ under F; that is, $F([a, b])$. We may think of this curve as "connecting" the two points $F(a) = (f(a), g(a), h(a))$ and $F(b) = (f(b), g(b), h(b))$. Given two points $A, B \in R^3$, the question of whether or not there is a curve "connecting" A and B is therefore seen to be the same as the question of whether or not there is a continuous function $F:[a, b] \to R^3$ such that $F(a) = A$ and $F(b) = B$. Furthermore, the interval $[a, b]$ may just as well be restricted to $[0, 1]$, for using any homeomorphism $\varphi:[0, 1] \to [a, b]$, one may show that the required $F:[a, b] \to R^3$ exists if and only if a corresponding $G = F\varphi:[0, 1] \to R^3$ exists. These observations motivate the following two definitions:

DEFINITION 6.1 Let X be a topological space. A continuous function $f:[0, 1] \to X$ is called a *path* in X. The path f is said to *connect* or *join* the point $f(0)$ to the point $f(1)$. $f(0)$ is called the *initial point* and $f(1)$ is called the *terminal point* of the path f.

If f is a path in X, $f([0, 1])$ is called a *curve* in X.

DEFINITION 6.2 A topological space X is said to be *path-connected* if, for each pair of points $u, v \in X$, there is a path f connecting u to v.

A non-empty subset A of a topological space X is said to be *path-connected* if the topological space A in the relative topology is path-connected.

The real line R is a path-connected space, for if a, b are two real numbers, the path $f:[0, 1] \to R$ defined by

$$f(t) = a + (b - a)t$$

for $t \in [0, 1]$ connects $f(0) = a$ and $f(1) = b$. R^n is also path-connected. This may be seen by either joining a given pair x, y of points of R^n by a path, or by using the general result that if X and Y are path-connected spaces, then so is $X \times Y$ (see Exercise 5 of this section). Another significant class of path-connected spaces is the spheres, S^n, $n > 0$.

A path f in a topological space X whose initial and terminal points coincide is called a *closed* path or a *loop* in X. Though such paths play a significant role in topology, we shall not be concerned with them in this section.

If f is a path in a topological space X and g is a continuous mapping of X into a second topological space Y, then the composite function $gf:[0, 1] \to Y$ is a path in Y.

THEOREM 6.3 Let Y be a topological space. If there exists a path-connected topological space X and a continuous mapping $g:X \to Y$, which is onto, then Y is path-connected.

Proof. Let $a, b \in Y$. Since $g:X \to Y$ is onto, there are points $a', b' \in X$ such that $g(a') = a$, $g(b') = b$. Since X is path-connected, there is a path f in X joining a' to b' and, consequently, the path gf joins a to b.

Note the necessity of the requirement that $g:X \to Y$ be onto. It follows that given homeomorphic topological spaces X and Y, X is path-connected if and only if Y is path-connected. Thus, path-connectedness is a topological property.

Path-connectedness is a stronger property than connectedness; that is, if a topological space X is path-connected then X is connected.

THEOREM 6.4 Let X be a path-connected topological space, then X is connected.

Proof. Suppose X were not connected. Then there is a proper subset P of X which is both open and closed. Since P is proper, there is a point $a \in P$ and a point $b \in C(P)$. Let $f:[0, 1] \to X$ be a path from a to b. $f^{-1}(P)$ is a proper subset of $[0, 1]$ for $0 \in f^{-1}(P)$, $1 \notin f^{-1}(P)$. Since f is continuous, $f^{-1}(P)$ is both open and closed. But this contradicts the fact that $[0, 1]$ is connected. Therefore X is connected.

The converse of Theorem 6.4 is false. A counter-example to the converse, that is, a topological space that is connected but not path-connected, is the subspace of the plane consisting of the set of points (x, y) such that either

$$x = 0, \quad -1 \leqq y \leqq 1,$$

or

$$0 < x \leqq 1, \quad y = \cos \frac{\pi}{x}.$$

One may obtain some idea of this space by referring to Figure 15, where we have tried to show the main characteristics of this space. It is impossible to picture this space completely, for, as the values

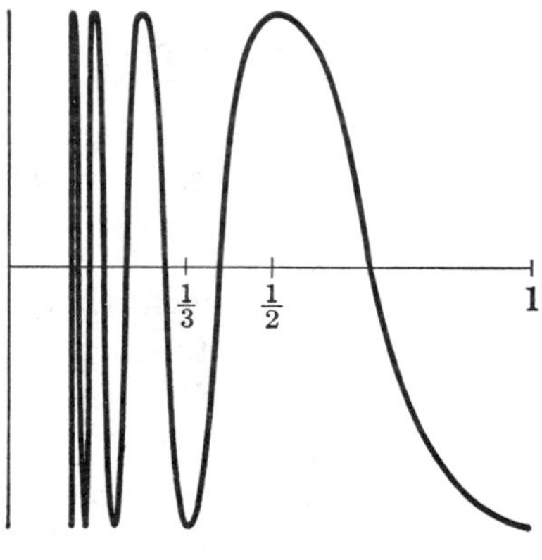

Figure 16

of x approach 0, the oscillation of the graph $y = \cos \frac{\pi}{x}$ becomes more and more rapid.

It is not difficult to prove that this space is connected. First of all let us decompose this space into two subsets Z_1 and Z_2, where Z_1 is the set of points $(0, y)$, $-1 \leqq y \leqq 1$, on the Y-axis, and Z_2 is the complementary set consisting of those points (x, y), $0 < x \leqq 1$ and $y = \cos \frac{\pi}{x}$. The function $F(t) = \left(t, \cos \frac{\pi}{t}\right)$ defines a continuous mapping of the connected interval $(0, 1]$ onto Z_2, hence Z_2 is connected. To prove that the entire space $Z = Z_1 \cup Z_2$ is connected, we shall prove that $\overline{Z}_2 = Z$; that is, $Z_1 \subset \overline{Z}_2$. This is so because there are points of Z_2 arbitrarily close to each point of Z_1. For, let $(0, y) \in Z_1$ and let $\varepsilon > 0$ be given. We may find an even integer N sufficiently large so that $\frac{1}{N} < \varepsilon$. Now $\cos \frac{\pi}{1/N} = 1$ and $\cos \frac{\pi}{1/(N+1)} = -1$, hence by the intermediate-value theorem there is a number $t \in \left[\frac{1}{N+1}, \frac{1}{N}\right]$ such that $\cos \frac{\pi}{t} = y$. The point $\left(t, \cos \frac{\pi}{t}\right)$ is in Z_2 and its distance from $(0, y)$ is less than ε. Thus $Z_1 \subset \overline{Z}_2$ and $\overline{Z}_2$ is the entire space Z. By Corollary 5.6, Z is connected.

Now suppose there was a path $F:[0, 1] \to Z$ with initial point $F(0) = (0, 1) \in Z_1$ and terminal point $F(1) = (1, -1) \in Z_2$. Let us write $F(t) = (F_1(t), F_2(t))$. Then F_1 and F_2 are continuous functions and $F_1(0) = 0$, $F_1(1) = 1$. The set $U = F_1^{-1}(\{0\})$ is a closed bounded subset of the real numbers and hence contains its least upper bound t^*. Since $F_1(1) \neq 0$, $t^* < 1$. We shall show that because of the oscillation of $\cos \frac{\pi}{x}$ for values of x close to zero, the function F_2 cannot be continuous at t^*. For each value of t such that $t^* < t \leqq 1$ we have $F_1(t) > 0$, hence $F(t) \in Z_2$ and $F_2(t) = \cos \frac{\pi}{F_1(t)}$. We shall show that for each $\delta > 0$ with $t^* + \delta \leqq 1$, there is a value of t such that $|t^* - t| < \delta$ whereas $|F_2(t^*) - F_2(t)| \geqq 1$. First $F_1(t^* + \delta) > 0$, hence there is an even integer N sufficiently large so that $F_1(t^*) = 0 < \frac{1}{N+1} < \frac{1}{N} < F_1(t^* + \delta)$. By the intermediate-value theorem, we may find $u, v \in [t^*, t^* + \delta]$ such that $F_1(u) = \frac{1}{N+1}$, $F_1(v) = \frac{1}{N}$. Since $u, v > t^*$ we have $F_2(u) = \cos \frac{\pi}{F_1(u)} = \cos (N + 1)\pi = -1$, $F_2(v) = \cos \frac{\pi}{F_1(v)} = \cos N\pi = 1$. Thus, if $F_2(t^*) \geqq 0$, $|F_2(t^*) - F_2(u)| \geqq$

$1 \geqq \varepsilon$, whereas if $F_2(t^*) \leqq 0$, $|F_2(t^*) - F_2(v)| \geqq 1 \geqq \varepsilon$. This contradicts the continuity of F_2 at t^*. Thus no path such as F exists and therefore our space Z is not path-connected.

EXERCISES

1. Prove directly by constructing appropriate paths that the topological spaces R^n, I^n (the unit cube), and $S^n (n > 0)$ are path-connected.
2. Verify that in a topological space X

   (i) if there is a path with initial point A and terminal point B, then there is a path with initial point B and terminal point A, and

   (ii) if there is a path connecting points A and B and a path connecting points B and C, then there is a path connecting points A and C.
3. The *path component* of a point x in a topological space X is the set of all points of X that may be connected to x by a path in X. Denote this subset by $P\text{Cmp}(x)$. Verify:

   (i) for each $x \in X$, $x \in P\text{Cmp}(x)$;

   (ii) for each $x, y \in X$, if $y \in P\text{Cmp}(x)$, then $x \in P\text{Cmp}(y)$;

   (iii) for each $x, y, z \in X$, if $y \in P\text{Cmp}(x)$ and $z \in P\text{Cmp}(y)$, then $z \in P\text{Cmp}(x)$;

   (iv) for each $x \in X$, $P\text{Cmp}(x)$ is path-connected;

   (v) if A is a path-connected subset of X, then $A \subset P\text{Cmp}(x)$ for some $x \in X$.

   (vi) X is path-connected if and only if $X = P\text{Cmp}(x)$ for some $x \in X$.
4. If A and B are path-connected subsets of a topological space X and $A \cap B \neq \emptyset$, then $A \cup B$ is path-connected.
5. Let $\{X_\alpha\}_{\alpha \in A}$ be an indexed family of topological spaces and set $X = \Pi_{\alpha \in A} X_\alpha$. For each $\alpha \in A$ let $f_\alpha : I \to X_\alpha$ be a path in X_α. Set $(f_A(t))(\alpha) = f_\alpha(t)$ so that $f_A : I \to X$. Prove that f_A is a path in X. Prove that if each X_α is path-connected, so is X.
6. Let X be a topological space, and let TX and SX be the cone over X and the suspension of X respectively. Prove that TX and SX are both path-connected.

7 HOMOTOPIC PATHS AND THE FUNDAMENTAL GROUP

The collection of points on and between the two concentric circles $x^2 + y^2 = 1$ and $x^2 + y^2 = 2$ is called an *annulus*. It is easy to see that this annulus is path-connected. For example, given two points $p_0 = (x_0, y_0)$ and $p_1 = (x_1, y_1)$, one may construct a path from p_0 to p_1 by first traversing the radius on which p_0 lies until we reach a point whose distance from the origin is the same as that of p_1 and then traversing in a clockwise direction the circular arc from this point to p_1 (see Figure 17). Let us call this path F_0. Alternately, one may construct a second path F_1 from p_0 to p_1,

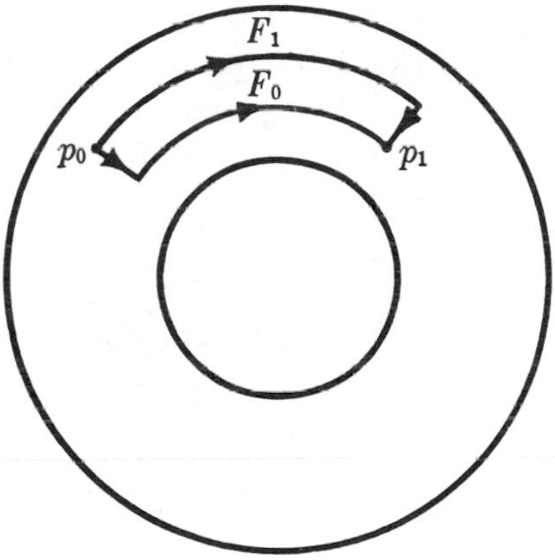

Figure 17

by first traversing in a clockwise direction a circular arc from p_0 to the radius on which p_1 lies and then traversing this radius until p_1 is reached. If, for the moment, we think of each of these two paths F_0 and F_1 as being represented by elastic strings with initial point p_0 and terminal point p_1, it is clear that in a given unit of time it would be possible to smoothly deform the path F_0 into the path F_1 (keeping p_0 and p_1 fixed throughout the deformation). This deformation might be carried out so that at time $t = \frac{1}{4}$

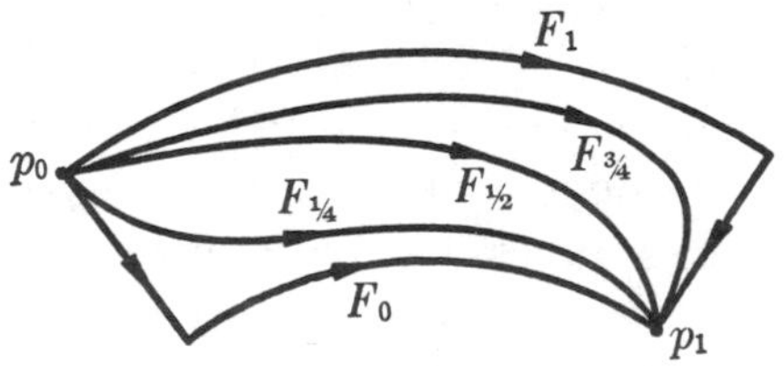

Figure 18

the string lies over the curve $F_{1/4}$ of Figure 18, at time $t = \frac{1}{2}$ the string lies over $F_{1/2}$, and at time $t = \frac{3}{4}$ the string lies over $F_{3/4}$. We may thus conceive of the deformation of the path F_0 into the path F_1 as being accomplished by constructing an entire family of paths F_t for $0 \leqq t \leqq 1$, such that if t and t' are close then the paths F_t and $F_{t'}$ are "close."

The concept of regarding two paths as being "close" implies the introduction of some sort of topology in this set of paths. Although this topology might be introduced directly by defining open set or neighborhood in the set of paths, an easier procedure is first to regard the unit of time as a unit interval on a line. Instead of viewing the two original paths F_0 and F_1 as being defined on the same unit interval, let us view F_0 as temporarily being defined on the homeomorphic image I_0 of the unit interval, where I_0 is the set of points $(x, 0)$ in the plane with $0 \leqq x \leqq 1$ (see Figure 19). Similarly, let us view F_1 as being defined on I_1, where I_1 is the set of points $(x, 1)$, $0 \leqq x \leqq 1$. For each value of t, $0 \leqq t \leqq 1$, we may view the path F_t as being defined on the homeomorphic image of the unit interval I_t, where I_t is the set of points (x, t), $0 \leqq x \leqq 1$. If we have such a situation, we may define a function $H: I^2 \to X$, where I^2 is the unit square and X is our annulus, by setting $H(x, t) = F_t(x, t)$, as depicted in Figure 19. Equivalently, if we insist on viewing each path F_t as

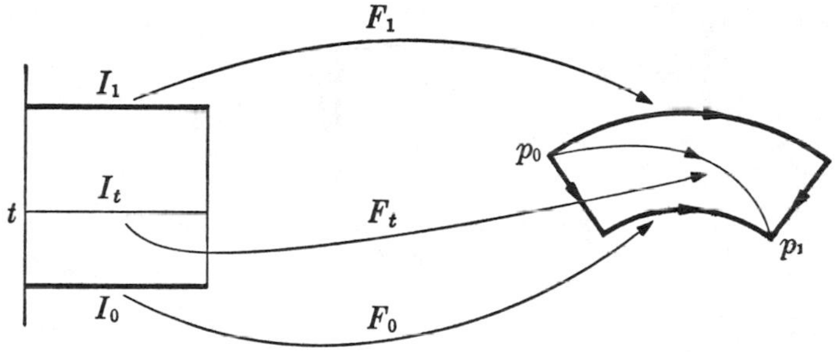

Figure 19

being defined on the same unit interval I, we may still obtain the same function H by setting $H(x, t) = F_t(x)$. We now introduce the concept of closeness amongst paths by requiring that the function $H: I^2 \to X$ be continuous.

DEFINITION 7.1 Let F_0, F_1 be two paths in a topological space X with the same initial point $p_0 = F_0(0) = F_1(0)$ and the same terminal point $p_1 = F_0(1) = F_1(1)$. F_0 is said to be *homotopic* to F_1 if there is a continuous function $H: I^2 \to X$ such that

$$\begin{aligned} H(0, t) &= p_0, & 0 \leqq t \leqq 1, \\ H(1, t) &= p_1, & 0 \leqq t \leqq 1, \\ H(x, 0) &= F_0(x), & 0 \leqq x \leqq 1, \\ H(x, 1) &= F_1(x), & 0 \leqq x \leqq 1. \end{aligned}$$

The function H is called a *homotopy connecting* F_0 to F_1.

In this event we say that the path F_0 is deformable into the path F_1 with fixed end points. One may illustrate the fact that a path F_0 is homotopic to F_1 by indicating that I^2 is the domain of the homotopy H, where the boundary of I^2 is mapped in agreement with the conditions of Definition 7.1 (see Figure 20).

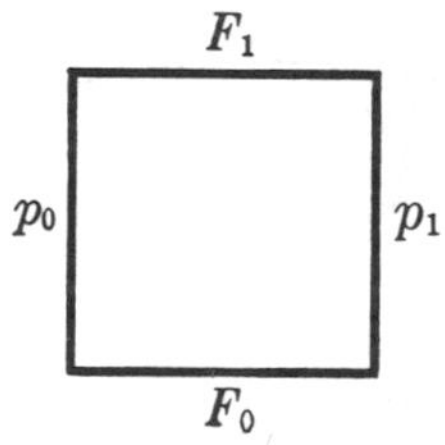

Figure 20

The relation of homotopy between paths satisfies the following three properties.

THEOREM 7.2 Let F_0, F_1, F_2 be three paths in a topological space X with the same initial point p_0 and the same terminal point p_1.

(i) F_0 is homotopic to itself.

(ii) If F_0 is homotopic to F_1 then F_1 is homotopic to F_0.

(iii) If F_0 is homotopic to F_1 and F_1 is homotopic to F_2 then F_0 is homotopic to F_2.

Proof. To show that F_0 is homotopic to itself we need only define $H: I^2 \to X$ by $H(x, t) = F_0(x)$. Next, suppose that F_0 is homotopic to F_1 so that there is a homotopy $H: I^2 \to X$ from F_0 to F_1. For each $(x, t) \in I^2$, set $H'(x, t) = H(x, 1 - t)$. Then H' is easily seen to be a homotopy from F_1 to F_0. To prove (iii), first let G be a homotopy from F_0 to F_1 and let H be a homotopy from F_1 to F_2. We may construct a homotopy from F_0 to F_2 in stages. First, we alter H to a function H' defined for (x, t') with $1 \leqq t' \leqq 2$ so that G and H' together constitute a function K' defined for (x, t) with $0 \leqq t \leqq 2$. Finally, we compress K' to a function K again defined

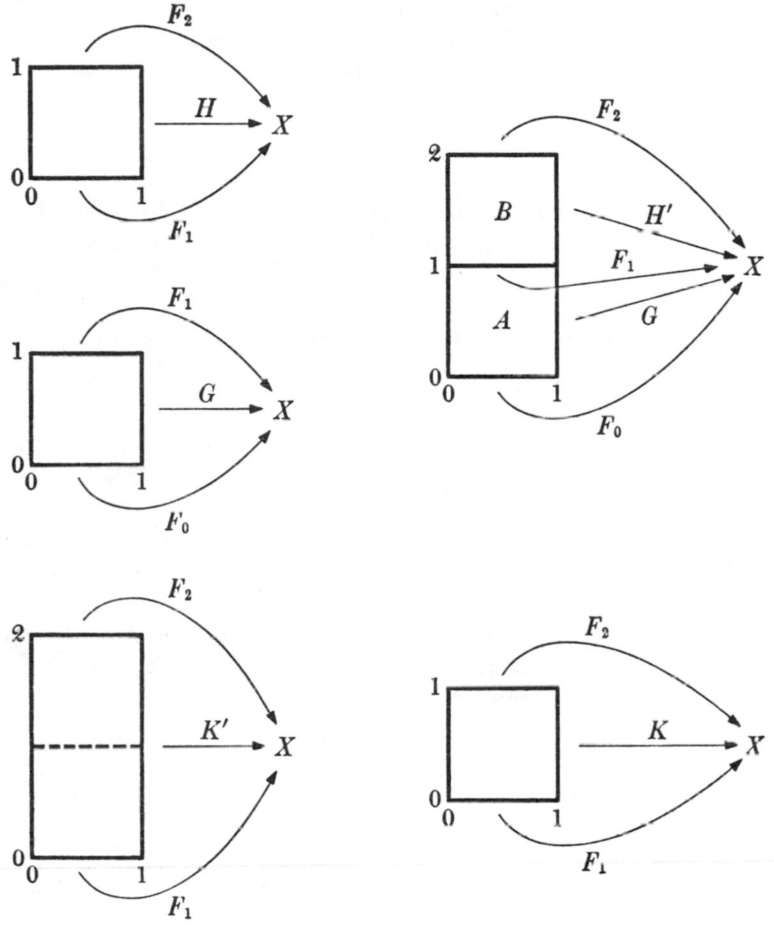

Figure 21

on I^2. The diagrams of Figure 21 depict this process. To this end let $H'(x, t') = H(x, t' - 1)$, $0 \leqq x \leqq 1$, $1 \leqq t' \leqq 2$. We then have two functions G and H', G defined on the subset $A = I^2$ of the plane and H' defined on the subset B consisting of the points (x, t) such that $0 \leqq x \leqq 1$ and $1 \leqq t \leqq 2$. The set $A \cap B$ consists of the points $(x, 1)$, $0 \leqq x \leqq 1$ and therefore we have $G(x, 1) = F_1(x)$, $H'(x, 1) = H(x, 0) = F_1(x)$; that is, G and H' agree in their common domain of definition. We shall now prove a lemma that asserts that together G and H' define a continuous function $K': A \cup B \to X$.

LEMMA 7.3 Let A, B be closed subsets of a topological space Z. Let $g: A \to X$ and $h: B \to X$ be continuous functions with the property that for $z \in A \cap B$, $g(z) = h(z)$. Then the function $k: A \cup B \to X$ defined by $k(z) = g(z)$, $z \in A$, $k(z) = h(z)$, $z \in B$, is a continuous extension of g and h.

Proof. Let U be a closed subset of X. Then $g^{-1}(U)$ is a relatively closed subset of A and, since A is closed, $g^{-1}(U)$ is a closed subset of Z. Similarly, $h^{-1}(U)$ is a closed subset of Z. But $k^{-1}(U) = g^{-1}(U) \cup h^{-1}(U)$, hence $k^{-1}(U)$ is closed and k is continuous.

Continuing now with the proof of Theorem 7.2, the function $K': A \cup B \to X$ defined by $K'(x, t) = G(x, t)$, $(x, t) \in A$, $K'(x, t) = H'(x, t)$, $(x, t) \in B$ is continuous. We finally "compress" K' to the function $K: I^2 \to X$ defined by $K(x, t) = K'(x, 2t)$, $(x, t) \in I^2$. To recapitulate, for $(x, t) \in I^2$ with $0 \leqq t \leqq \frac{1}{2}$, we have

$$K(x, t) = K'(x, 2t) = G(x, 2t), \quad 0 \leqq t \leqq \tfrac{1}{2},$$

whereas for $\frac{1}{2} \leqq t \leqq 1$, we have

$$K(x, t) = K'(x, 2t) = H'(x, 2t) = H(x, 2t - 1), \quad \tfrac{1}{2} \leqq t \leqq 1.$$

From these two equations it follows that $K(0, t)$ is the initial point of F_0 and F_2, $K(1, t)$ is the terminal point of F_0 and F_2, and that $K(x, 0) = G(x, 0) = F_0(x)$, whereas $K(x, 1) = H(x, 1) = F_2(x)$. Therefore K is a homotopy from F_0 to F_2. This completes the proof of Theorem 7.2.

If F is a path that is homotopic to a path G we shall write $F \cong G$. Theorem 7.2 then states: (i) $F_0 \cong F_0$; (ii) if $F_0 \cong F_1$ then $F_1 \cong F_0$; (iii) if $F_0 \cong F_1$ and $F_1 \cong F_2$ then $F_0 \cong F_2$. Thus $\cong$ is an equivalence relation. We shall denote the equivalence class of a path F by $[\![F]\!]$.

Definition 7.4 An equivalence set of homotopic paths is called a *homotopy class of paths.* At a point z in a topological space Z the collection of homotopy classes of closed paths at z is denoted by $\Pi(Z, z)$. Among these homotopy classes there is the homotopy class $[\![e_z]\!]$, where e_z is the *constant* path defined by $e_z(t) = z$, $0 \leqq t \leqq 1$.

The remainder of this section will be devoted to showing that there is a natural procedure whereby $\Pi(Z, z)$ may be converted into a group with $[\![e_z]\!]$ as its identity.

Definition 7.5 Let $F, G: I \to Z$ be closed paths at $z \in Z$. Define $F \cdot G: I \to Z$ by

$$(F \cdot G)(t) = F(2t), \qquad 0 \leqq t \leqq \tfrac{1}{2},$$
$$(F \cdot G)(t) = G(2t - 1), \quad \tfrac{1}{2} \leqq t \leqq 1.$$

Since $F(1) = G(0) = z$, by Lemma 7.3 $F \cdot G$ is a closed path at z. $F \cdot G$ is called the *product* or *concatenation* of F and G or F *followed by* G. We will now show that this product induces a product in $\Pi(Z, z)$.

Lemma 7.6 In $\Pi(Z, z)$ let $[\![F]\!] = [\![F']\!]$ and $[\![G]\!] = [\![G']\!]$, then $[\![F \cdot G]\!] = [\![F' \cdot G']\!]$.

Proof. We are given homotopies $K, L: I^2 \to Z$ connecting F to F' and G to G' respectively. In effect, a concatenation of the homotopies K and L yields a homotopy connecting $F \cdot G$ to $F' \cdot G'$. Let $H(t, s) = K(2t, s)$, $0 \leqq t \leqq \frac{1}{2}$, and $H(t, s) = L(2t - 1, s)$, $\frac{1}{2} \leqq t \leqq 1$. Since $K(1, s) = L(0, s) = z$, Lemma 7.3 shows that H is continuous, while $H(t, 0) = (F \cdot G)(t)$, $H(t, 1) = (F' \cdot G')(t)$, and $H(0, s) = H(1, s) = z$.

Definition 7.7 In $\Pi(Z, z)$ let $[\![F]\!] \cdot [\![G]\!] = [\![F \cdot G]\!]$.

Lemma 7.8 $[\![F]\!] \cdot [\![e_z]\!] = [\![e_z]\!] \cdot [\![F]\!] = [\![F]\!]$ for all $[\![F]\!] \in \Pi(Z, z)$.

Proof. We shall first show that $[\![F]\!] \cdot [\![e_z]\!] = [\![F]\!]$. Define $H: I^2 \to Z$ by

$$H(t, s) = F\left(\frac{2t}{s+1}\right), \quad s \geqq 2t - 1,$$

$$H(t, s) = z, \qquad s \leqq 2t - 1.$$

If $(t, s) \in I^2$ and $s = 2t - 1$, then $\dfrac{2t}{s+1} = 1$. Thus, by Lemma 7.3, H is continuous. If $s = 1$ then $s \geqq 2t - 1$ and $H(t, 1) = F(t)$. If $s = 0$ then for $0 \leqq t \leqq \frac{1}{2}$ we have $s \geqq 2t - 1$ so that $H(t, 0) = F(2t)$ while for $\frac{1}{2} \leqq t \leqq 1$ we have $2t - 1 \geqq s$ so that $H(t, 0) = e_z(t) = z$. Therefore $H(t, 0) = (F \cdot e_z)(t)$ and H connects $F \cdot e_z$ to F.

To show that $[\![e_z]\!] \cdot [\![F]\!] = [\![F]\!]$, we define $H(t, s) = z$ for $s \geqq 2t$ and $H(t, s) = F\left(\dfrac{2t - s}{2 - s}\right)$ for $s \leqq 2t$ and in similar fashion show that H connects F_z to $e_z \cdot F$.

The apparent complexity of the expressions for the homotopies H in these two cases is explained by the use of the following two Figures. In Figure 22 we have projected the point $(t, s) \in I^2$ with $s \geqq 2t - 1$ onto the point $(t', 1) \in I^2$ from the point $(0, -1)$. In this fashion as s goes from 0 to 1 the interval $(t, 0)$, $0 \leqq t \leqq \frac{1}{2}$ is gradually enlarged until it becomes the interval $(t', 1)$, $0 \leqq t' \leqq 1$. By analytic geometry $t' = \dfrac{2t}{s+1}$. By setting $H(t, s) = F\left(\dfrac{2t}{s+1}\right)$ we have arranged matters so that for a fixed s the interval (t, s), $0 \leqq t \leqq \dfrac{s+1}{2}$ is mapped in such a way as to trace out the same path as F. Finally the interval $\dfrac{s+1}{2} \leqq t \leqq 1$ is mapped into z. Thus we have started out along the interval $(t, 0)$, $0 \leqq t \leqq 1$, mapping this horizontal interval by $F \cdot e_z$ and gradually, as s increases, increased the length of the horizontal interval mapped using F and decreased to zero the length of the horizontal interval mapped by e_z.

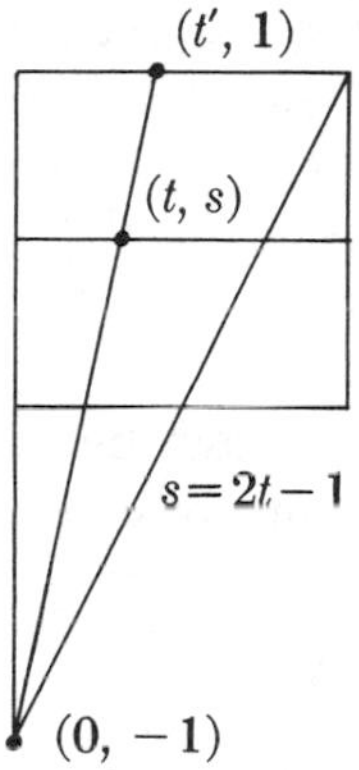

Figure 22

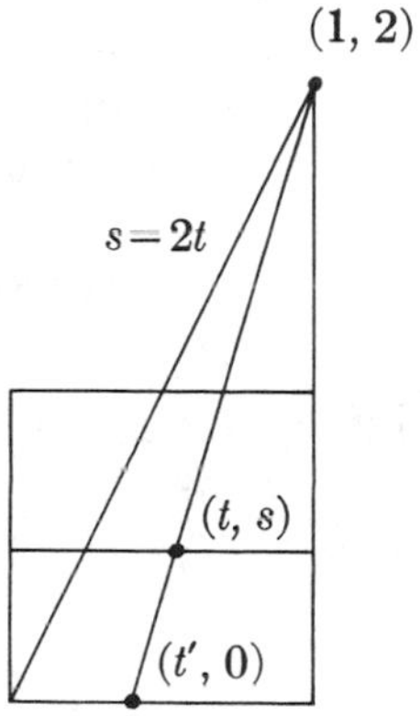

Figure 23

Similarly, using Figure 23, the projection of the point $(t, s) \in I^2$ with $s \leqq 2t$ onto the point $(t', 0) = \left(\frac{2t - s}{2 - s}, 0\right)$ from the point $(1, 2)$ results in gradually contracting the interval $(t, 0)$, $0 \leqq t \leqq 1$, into the interval $(t', 1)$, $\frac{1}{2} \leqq t' \leqq 1$.

DEFINITION 7.9 Let $F: I \to Z$ be a path. Define $F^{-1}: I \to Z$ by $F^{-1}(t) = F(1 - t)$.

If F is a path from z to y then F^{-1} is a path from y to z. In particular if F is a closed path at z then F^{-1} is also a closed path at z which may be thought of as F traversed in the opposite sense.

LEMMA 7.10 For each $[\![F]\!] \in \Pi(Z, z)$, $[\![F]\!] \cdot [\![F^{-1}]\!] = [\![F^{-1}]\!] \cdot [\![F]\!] = [\![e_z]\!]$.

Proof. We must show that $F \cdot F^{-1} \simeq e_z \simeq F^{-1} \cdot F$. To show that $F \cdot F^{-1} \simeq e_z$ define $H: I^2 \to Z$ as follows:

$$\begin{aligned} H(t, s) &= F(2t), & & s \geqq 2t; \\ H(t, s) &= F(s), & & s \leqq 2t \text{ and } s \leqq -2t + 2; \\ H(t, s) &= F(2 - 2t), & & s \geqq -2t + 2. \end{aligned}$$

Since the various definitions of H agree when $s = 2t$ and $s = -2t + 2$, H is continuous. By setting $s = 0$ and $s = 1$, H is easily seen to be a homotopy connecting e_z to $F \cdot F^{-1}$. Interchanging the roles of F and F^{-1} yields a homotopy connecting e_z to $F^{-1} \cdot F$.

Figure 24 may be used to explain the construction of the homotopy H. We have marked the upper and lower edges of I^2

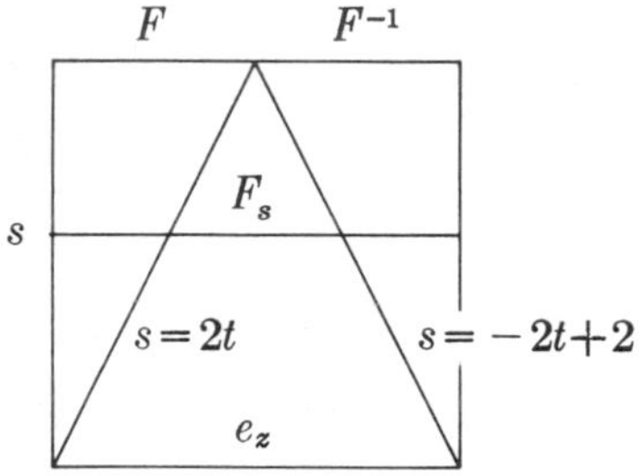

Figure 24

with the symbols for paths to indicate how the mapping H behaves along these edges. Let $F_s(t) = H(t, s)$. The path F_s starts out by tracing the path of F at twice its normal rate until $s = 2t$ whereupon it remains stationary at $F(s)$ until $s = -2t + 2$. The path F_s then returns to z backwards along this portion of the path of F, again at twice the normal rate.

We have thus shown that every element of $\Pi(Z, z)$ has an inverse. To complete the proof that $\Pi(Z, z)$ is a group we must show that the product is associative.

LEMMA 7.11 $([\![F]\!] \cdot [\![G]\!]) \cdot [\![K]\!] = [\![F]\!] \cdot ([\![G]\!] \cdot [\![K]\!])$ for all $[\![F]\!], [\![G]\!], [\![K]\!] \in \Pi(Z, z)$.

Proof. We must show that $(F \cdot G) \cdot K \cong F \cdot (G \cdot K)$. We define $H: I^2 \to Z$ as follows:

$$H(t, s) = F\left(\frac{4t}{s + 1}\right), \qquad 4t - 1 \leqq s;$$

$$H(t, s) = G(4t - s - 1), \qquad 4t - 2 \leqq s \leqq 4t - 1;$$

$$H(t, s) = K\left(\frac{4t - 2s}{2 - s} - 1\right), \quad s \leqq 4t - 2.$$

The various definitions of H agree when $s = 4t - 1$ and $s = 4t - 2$ so H is continuous. Again by setting $s = 0$ and $s = 1$ it is easily seen that H is a homotopy connecting $(F \cdot G) \cdot K$ with $F \cdot (G \cdot K)$.

In Figure 25 we have illustrated the homotopy H. A point (t, s) with $4t - 1 \leqq s$ is projected from $(0, -1)$ onto the point

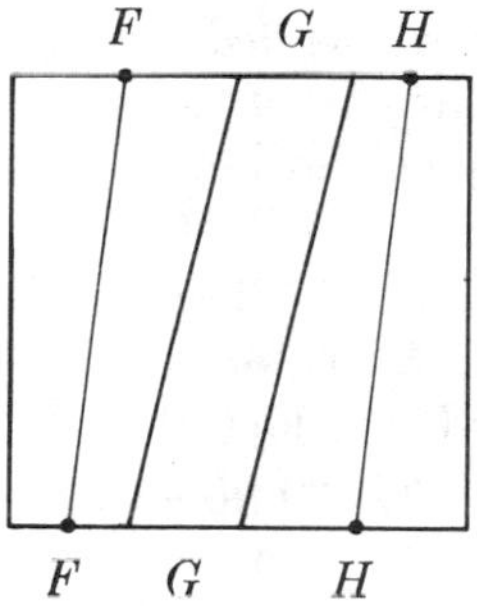

Figure 25

$\left(\frac{t}{s + 1}, 0\right)$ which by $(F \cdot G) \cdot K$ is mapped into $((F \cdot G) \cdot K)\left(\frac{t}{s + 1}\right)$ $= F\left(\frac{4t}{s + 1}\right)$. Points (t, s) with $4t - 2 \leqq s \leqq 4t - 1$ are parallel projected onto $\left(\frac{4t - s}{4}, 0\right)$ which by $(F \cdot G) \cdot K$ are mapped into $((F \cdot G) \cdot K)\left(\frac{4t - s}{4}\right) = G(4t - s - 1)$. Finally a point (t, x) with $s \leqq 4t - 2$ is projected from $(1, 2)$ onto the point $\left(\frac{2t - s}{2 - s}, 0\right)$ which by $(F \cdot G) \cdot K$ is mapped into $((F \cdot G) \cdot K)\left(\frac{2t - s}{2 - s}\right) =$ $K\left(\frac{4t - 2s}{2 - s} - 1\right)$.

EXERCISES

1. Let X, Y be topological spaces and $f: X \to Y$ be a continuous function with $f(x) = y$. Let g and g' be closed paths at $x \in X$. Prove that $fg \cong fg'$ whenever $g \cong g'$. Set $f_*[\![g]\!] = [\![fg]\!]$. Prove that f_* is a homomorphism from $\Pi(X, x)$ to $\Pi(Y, y)$.
2. The category of *topological spaces with base points* has as its objects pairs of the form (X, x) where X is a topological space and $x \in X$ and has as its mappings functions $f: (X, x) \to (Y, y)$ such that $f: X \to Y$ is a continuous function and $f(x) = y$. Set $F_1(X, x) = \Pi(X, x)$ and $F_2(f) = f_*$ as defined in Exercise 1. Prove that (F_1, F_2) is a functor from the category of topological spaces with base points to the category of groups and homomorphisms as defined in Section 9, Chapter 3.
3. Two groups G and G' are called *isomorphic* if there are homomorphisms $h: G \to G'$ and $h': G' \to G$ such that $h'h$ is the identity mapping of G and hh' is the identity mapping of G'. Prove that if $f: X \to Y$ is a homeomorphism of the topological space X with the space Y such that $f(x) = y$ then $\Pi(X, x)$ is isomorphic to $\Pi(Y, y)$.
4. Let A be a subspace of a topological space X. If there is a continuous function $r: X \to A$ such that $r(a) = a$ for all $a \in A$, A is called a *retract* of X and r is called a *retraction*, i.e., $ri = 1_A$ where $i: A \to X$ is the inclusion map and 1_A is the identity mapping of A. Let $a_0 \in A$. Prove that if $r: X \to A$ is a retraction then $i_*: \Pi(A, a_0) \to \Pi(X, a_0)$ is one-one and $r_*: \Pi(X, a_0) \to \Pi(A, a_0)$ is onto. Prove that a circle on the boundary of an annulus is a retract of the annulus.
5. Two continuous functions $f, g: X \to Y$ are said to be *homotopic* if there is a continuous function $H: X \times I \to Y$ such that $H(x, 0) = f(x)$, $H(x, 1) = g(x)$. If furthermore for some $x_0 \in X$ we have $H(x_0, s) = f(x_0) = g(x_0)$, $s \in I$, they are said to be *homotopic rel* x_0. Let $f, g: X \to Y$ be homotopic rel x_0 and let $p: I \to X$ be a closed path at x_0. Set $K(t, s) = H(p(t), s)$. Prove that K is a homotopy connecting fp to gp and that therefore $f_* = g_*$.
6. A subspace A of a topological space X is called a *deformation retract rel* x_0 of X if there is a retraction $r: X \to A$ such that $ir: X \to X$ is homotopic rel x_0 to the identity map of X where i is the inclusion

map. Prove that in this case $\Pi(A, a_0)$ and $\Pi(X, a_0)$ are isomorphic. Prove that the center x_0 of a closed disc is a deformation retract rel x_0 of the disc. Let C be the circular boundary of a closed disc D. Prove that if $\Pi(C, c)$ contains more than one element then C cannot be a retract of D.

8 SIMPLE CONNECTEDNESS

DEFINITION 8.1 A topological space Z is said to be *simply connected* if at each point $z \in Z$ there is only one homotopy class of closed paths.

Thus if Z is simply connected, at each point $z \in Z$ the fundamental group $\Pi(Z, z)$ consists of precisely the identity element

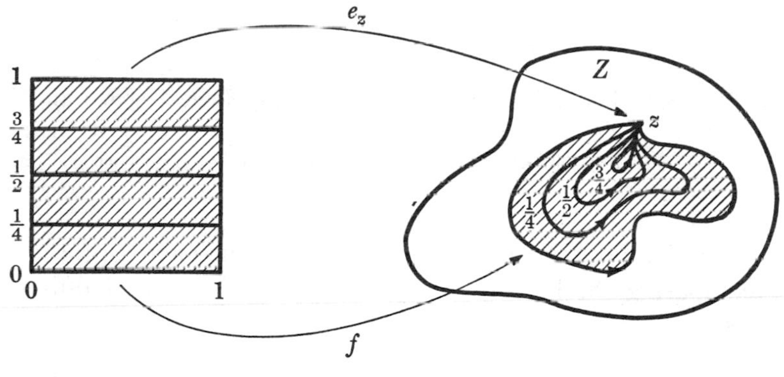

Figure 26

$[\![e_z]\!]$. In this case there is for each closed path f at z a homotopy $H: I^2 \to Z$ which deforms f into the constant path e_z, as depicted in Figure 26. The possibility of carrying out the deformation corresponds to the fact that the curve traced out by f does not enclose any holes in the space Z.

One can prove that an annulus is not simply connected, for, although a closed path such as C_1 (see Figure 27) is homotopic to a constant path, a closed path such as C_2 is not homotopic to a constant path.

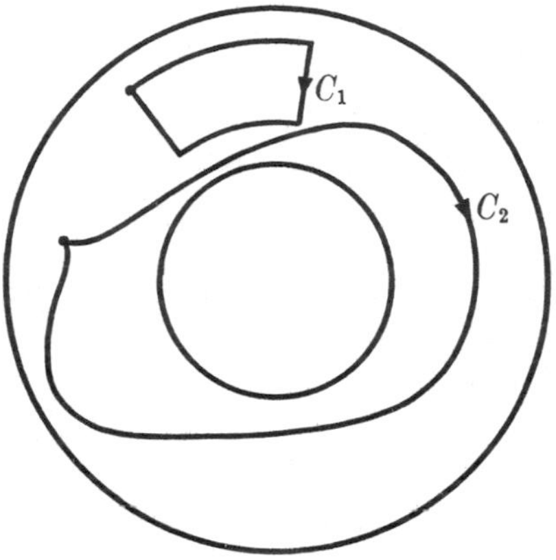

Figure 27

THEOREM 8.2 Let Z be a path-connected topological space and let $z \in Z$. Z is simply connected if and only if there is exactly one homotopy class of closed paths at z.

In order to prove this theorem we must develop a procedure for comparing the homotopy classes of closed paths at different points.

DEFINITION 8.3 Let f be a path in a topological space Z with $z = f(0)$ and $y = f(1)$. Let g be a closed path at y. Define $g_f: I \to Z$ by

$$\begin{aligned} g_f(t) &= f(3t), & 0 \leqq t \leqq \tfrac{1}{3}, \\ g_f(t) &= g(3t - 1), & \tfrac{1}{3} \leqq t \leqq \tfrac{2}{3}, \\ g_f(t) &= f(3 - 3t), & \tfrac{2}{3} \leqq t \leqq 1. \end{aligned}$$

g_f is a closed path at z which is constructed in accordance with Figure 28. In particular if g is the constant path e_y the

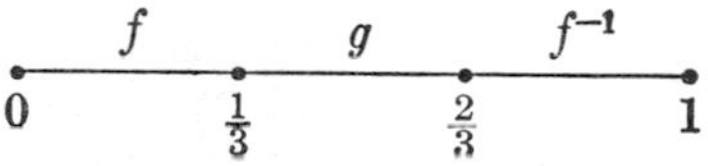

Figure 28

same homotopy used in the proof of Theorem 7.10 shows that $(e_y)_f \cong e_z$.

LEMMA 8.4 Let $[\![g]\!] = [\![g']\!] \in \Pi(Z, y)$, then $[\![g_f]\!] = [\![g'_f]\!] \in \Pi(Z, z)$.

Proof. Let $K: I^2 \to Z$ be the homotopy connecting g to g'. Define $H: I^2 \to Z$ as follows:

$$\begin{aligned} H(t, s) &= f(3t), & 0 \leqq t \leqq \tfrac{1}{3}; \\ H(t, s) &= K(3t - 1, s), & \tfrac{1}{3} \leqq t \leqq \tfrac{2}{3}; \\ H(t, s) &= f(3 - 3t), & \tfrac{2}{3} \leqq t \leqq 1. \end{aligned}$$

In the usual fashion one verifies that H is a homotopy connecting g_f to g'_f.

We may picture the homotopy H as being constructed in accordance with Figure 29. The homotopy K has been contracted by a factor of 3 on the t-axis so that it can occupy the middle strip of I^2, while the first and third segments of each horizontal line are appropriately contracted repetitions of f and f^{-1}. In fact, if f' is homotopic to f it is easily seen that using a contraction of this homotopy to map the first strip and the reverse to map the third strip we can obtain a homotopy connecting g_f and $g_{f'}$. We have thus shown:

LEMMA 8.5 Let f and f' be homotopic paths with $f(0) = f'(0) = z$ and $f(1) = f'(1) = y$. Then for $[\![g]\!] \in \Pi(Z, y)$, $[\![g_f]\!] = [\![g_{f'}]\!]$.

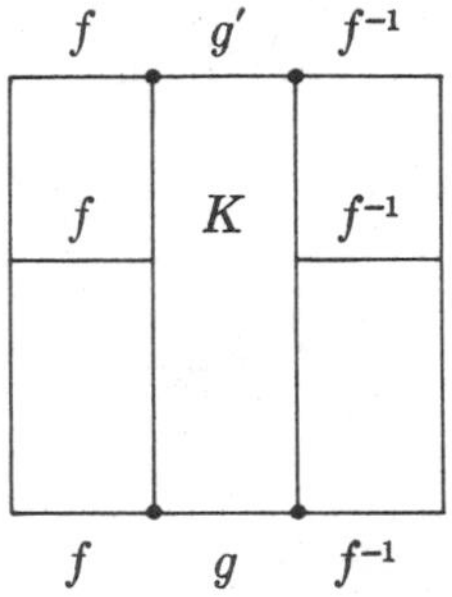

Figure 29

DEFINITION 8.6 Let $f:I \to Z$ be a path with $z = f(0)$ and $y = f(1)$. For $[\![g]\!] \in \Pi(Z, y)$ set $a_f([\![g]\!]) = [\![g_f]\!]$.

PROPOSITION 8.7 $a_f:\Pi(Z, y) \to \Pi(Z, z)$ is a homomorphism and if $f \cong f'$ then $a_f = a_{f'}$.

Proof. Since $(e_y)_f \cong e_z$, a_f carries the identity of $\Pi(Z, y)$ into that of $\Pi(Z, z)$. We must show that

$$a_f([\![g]\!]\cdot[\![h]\!]) = (a_f([\![g]\!]))\cdot(a_f([\![h]\!]))$$

for $[\![g]\!], [\![h]\!] \in \Pi(Z, y)$. Now $a_f([\![g]\!]\cdot[\![h]\!]) = a_f([\![g\cdot h]\!]) = [\![(g\cdot h)_f]\!]$ and $(a_f([\![g]\!]))\cdot(a_f([\![h]\!])) = [\![g_f]\!]\cdot[\![h_f]\!] = [\![g_f\cdot h_f]\!]$.

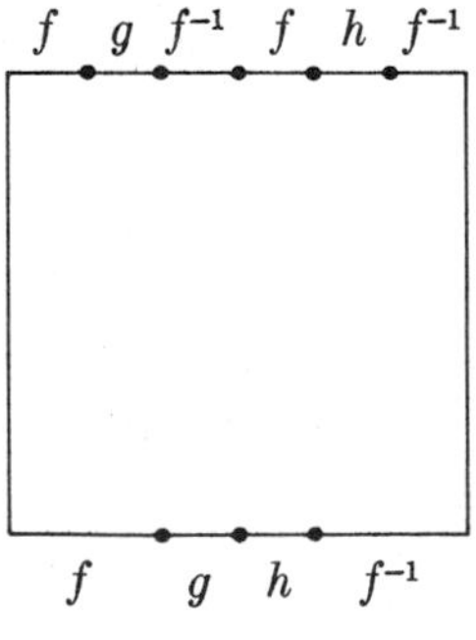

Figure 30

Thus we must show that $(g\cdot h)_f \cong g_f\cdot h_f$. In Figure 30 we have indicated how I^2 can be mapped along the lower edge by $(g \cdot h)_f$ and along the upper edge by $g_f \cdot h_f$. By now the procedure for constructing the appropriate homotopy should be clear. The last part follows from Lemma 8.5.

THEOREM 8.8 $a_f:\Pi(Z, y) \to \Pi(Z, z)$ and $a_{f^{-1}}:\Pi(Z, z) \to \Pi(Z, y)$ are inverse functions.

Proof. Suppose $[\![g]\!] \in \Pi(Z, y)$. Figure 31 shows how $(g_f)_{f^{-1}}$ is defined. Again a slight modification of the construction used in Lemma 7.10 provides a homotopy

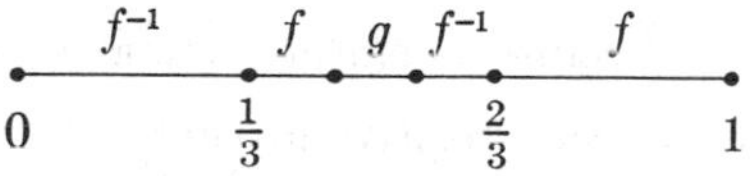

Figure 31

connecting $(g_f)_{f^{-1}}$ to g. Thus $a_{f^{-1}}(a_f([\![g]\!])) = [\![(g_f)_{f^{-1}}]\!] = [\![g]\!]$. Similarly $a_f a_{f^{-1}}$ is the identity.

If Z is path-connected and $\Pi(Z, y)$ consists of a single element $[\![e_y]\!]$, then for any point $z \in Z$ there is a path f from z to y and $\Pi(Z, z) = a_f(\Pi(Z, y))$ is also a single element. Thus Theorem 8.2 is a corollary to Theorem 8.8.

A homomorphism $a:G \to G'$ of a group G into a group G' which has an inverse is called an *isomorphism* and G and G' are said to be *isomorphic*. In this event a is one-one and onto. The relation G is isomorphic to G' is an equivalence relation. Theorem 8.8 therefore states that in a path-connected space the fundamental groups at any two points are isomorphic.

EXERCISES

1. An isomorphism of a group G with itself is called an *automorphism*. Let f and f' be paths in a space Z with $f(0) = f'(1) = z$ and

$f(1) = f'(0) = y$. Let $f' \cdot f^{-1}$ be the path defined by $(f' \cdot f^{-1})(t) = f'(2t)$, $0 \leqq t \leqq \frac{1}{2}$, $(f' \cdot f^{-1})(t) = f^{-1}(2t - 1)$, $\frac{1}{2} \leqq t \leqq 1$. Prove that $a_{f'}a_f$ is an automorphism of $\Pi(Z, y)$ such that $a_{f'}a_f([\![g]\!]) = [\![f' \cdot f^{-1}]\!] \cdot [\![g]\!] \cdot [\![f' \cdot f^{-1}]\!]^{-1}$.

2. The fundamental groupoid of a space Z has as its objects the points of Z and as its maps $H(z, y)$ the homotopy classes of paths from z to y. Define a rule of composition $H(z, y) \times H(y, w) \to H(z, w)$ so that the fundamental groupoid of Z becomes a category. Prove that for each $\alpha \in H(z, y)$ there is an element $\alpha^{-1} \in H(y, z)$ with $\alpha^{-1}\alpha = 1_z$ and $\alpha\alpha^{-1} = 1_y$. Let $f: Z \to W$ be continuous. Let $F_1(z) = f(z)$ and $F_2([\![g]\!]) = [\![fg]\!]$ for $[\![g]\!] \in H(z, y)$. Prove that (F_1, F_2) is a functor from the fundamental groupoid of Z to that of W.
3. Prove that a product of simply connected spaces is simply connected.
4. Prove that for each positive integer n, R^n and I^n are simply connected.

For further reading, in addition to the more general texts, Wall, *A Geometric Introduction to Topology*, Chinn and Steenrod, *First Concepts of Topology*, Massey, *Algebraic Topology: An Introduction*, and Wallace, *Introduction to Algebraic Topology* are highly recommended.

CHAPTER 5

Compactness

1 INTRODUCTION

A closed and bounded subset A of the real line R is characterized by the fact that for each collection $\{O_\alpha\}_{\alpha \in I}$ of open subsets of R such that $A \subset \bigcup_{\alpha \in I} O$, there is a finite subcollection $O_{\alpha_1}, O_{\alpha_2}, \ldots, O_{\alpha_n}$ with $A \subset \bigcup_{i=1}^{n} O_{\alpha_i}$. This second property is stated in terms that are applicable to any topological space. If this property holds in a particular topological space, the space is said to be "compact." The closed and bounded subsets of R^n are precisely the compact subspaces of R^n. This fact can be either proved directly or established by proving that the product of two compact spaces is itself compact. In metrizable spaces there is an alternate formulation of compactness; namely, that each infinite subset has a "point of accumulation."

Compactness, like connectedness and arcwise connectedness, is a "global" property, in that it depends on the nature of the entire space. The advantage in compact spaces is that one may study the whole space by studying a finite number of open subsets. We shall see this when we prove that a continuous function $f:X \to Y$ from a compact metric space X to a metric space Y is "uniformly continuous." In conclusion we shall examine some compact surfaces that may be formed by "identifying" edges of a rectangle.

2 COMPACT TOPOLOGICAL SPACES

DEFINITION 2.1 Let X be a set, B a subset of X, and $\{A_\alpha\}_{\alpha \in I}$ an indexed family of subsets of X. The collection $\{A_\alpha\}_{\alpha \in I}$ is called a *covering* of B or is said to *cover* B if $B \subset \bigcup_{\alpha \in I} A_\alpha$. If, in addition, the indexing set I is finite, $\{A_\alpha\}_{\alpha \in I}$ is called a *finite covering* of B.

Let X be a topological space and for each $x \in X$ let N_x be a neighborhood of x. Then $\{N_x\}_{x \in X}$ is a covering of X. For each integer n, let $A_n = [n, n + 1]$. Then $\{A_n\}_{n \in Z}$, where Z is the set of integers, is a covering of the set R of real numbers. Similarly, if for each ordered pair (m, n) of integers we let $A_{m,n}$ be the set of points $(x_1, x_2) \in R^2$ such that $m \leqq x_1 \leqq m + 1$, $n \leqq x_2 \leqq n + 1$, then $\{A_{m,n}\}_{(m,n) \in Z \times Z}$ is a covering of R^2. As a final example of a covering, let $X = R$ and let $B = (0, 1]$. If we set $A_1 = (\frac{1}{2}, 2)$, $A_2 = (\frac{1}{3}, 1)$, $A_3 = (\frac{1}{4}, \frac{1}{2})$, and in general, for each positive integer $n > 1$, set $A_n = \left(\dfrac{1}{n+1}, \dfrac{1}{n-1}\right)$, then $\{A_n\}_{n \in N}$, where N is the set of natural numbers, is a covering of B.

DEFINITION 2.2 Let X be a set and let $\{A_\alpha\}_{\alpha \in I}$, $\{B_\beta\}_{\beta \in J}$ be two coverings of a subset C of X. If for each $\alpha \in I$, $A_\alpha = B_\beta$ for some $\beta \in J$, then the covering $\{A_\alpha\}_{\alpha \in I}$ is called a *subcovering* of the covering $\{B_\beta\}_{\beta \in J}$.

Thus $\{A_\alpha\}_{\alpha \in I}$ is a subcovering of $\{B_\beta\}_{\beta \in J}$ if "every A_α is a B_β." In particular, if $\{B_\beta\}_{\beta \in J}$ is a covering of a subset C, and I is a subset of J such that $\{B_\beta\}_{\beta \in I}$ is also a covering of C, then $\{B_\beta\}_{\beta \in I}$ is a subcovering of $\{B_\beta\}_{\beta \in J}$. Let Q be the set of rational numbers and for each $q \in Q$, set $B_q = [q, q + 1]$. Then $\{B_q\}_{q \in Q}$ is a covering of the real numbers R. If again we let Z be the set of integers and $A_n = [n, n + 1]$, then $\{A_n\}_{n \in Z}$ is a subcovering of $\{B_q\}_{q \in Q}$.

Suppose that $f: X \to Y$ is a continuous function from a topological space X to a metric space Y. Given $\varepsilon > 0$, the continuity of f gives rise to a covering of X in the following manner. For each $x \in X$, given this $\varepsilon > 0$, there is an open neighborhood U_x of x such that the images under f of all points of U_x are within ε of $f(x)$, or equivalently, $f(U_x) \subset B(f(x); \varepsilon)$. The family $\{U_x\}_{x \in X}$ of these subsets of X is clearly a covering of X. This covering has the additional property that it is composed of open sets. We shall, naturally, refer to such a covering as an "open" covering.

Definition 2.3 Let X be a topological space and B a subset of X. A covering $\{A_\alpha\}_{\alpha \in I}$ of B is said to be an *open covering* of B if for each $\alpha \in I$, A_α is an open subset of X.

Definition 2.4 A topological space X is said to be *compact* if for each open covering $\{U_\alpha\}_{\alpha \in I}$ of X there is a finite subcovering $\{U_\beta\}_{\beta \in J}$.

As an alternate definition of compactness we may use the criterion, X is compact if for each open covering $\{U_\alpha\}_{\alpha \in I}$ of X there is a finite subset of indices $\{\alpha_1, \alpha_2, \ldots, \alpha_n\}$ such that the collection $U_{\alpha_1}, U_{\alpha_2}, \ldots, U_{\alpha_n}$ covers X.

Definition 2.5 A subset C of a topological space X is said to be *compact*, if C is a compact topological space in the relative topology.

A topological space C may be a subspace of two distinct larger topological spaces X and Y. In this event the relative topology of C is the same whether we regard C as a subspace of

X or of Y, and, consequently, the assertion C is compact depends only on C and its topology. We may relate the compactness of a subspace C of a topological space X to the topology of X by means of the following theorem.

THEOREM 2.6 A subset C of a topological space X is compact if and only if for each open covering $\{U_\alpha\}_{\alpha \in I}$ of C, U_α open in X, there is a finite subcovering $U_{\alpha_1}, U_{\alpha_2}, \ldots, U_{\alpha_n}$ of C.

Proof. Let C be compact and let $\{U_\alpha\}_{\alpha \in I}$ be an open covering of C. Then $\{U_\alpha \cap C\}_{\alpha \in I}$ is a covering of C by relatively open sets. Thus there is a finite subcovering $\{U_\alpha \cap C\}_{\alpha \in J}$ and $\{U_\alpha\}_{\alpha \in J}$ covers C. Conversely, suppose that for each open covering $\{U_\alpha\}_{\alpha \in I}$ of C there is a finite subcovering. Let $\{V_\beta\}_{\beta \in I}$ be a covering of C by relatively open subsets of C. For each $\beta \in I$, $V_\beta = U_\beta \cap C$ where U_β is open in X. Thus $\{U_\beta\}_{\beta \in I}$ is an open covering of C. By our hypothesis there is a finite subcovering $U_{\beta_1}, U_{\beta_2}, \ldots, U_{\beta_m}$. It follows that $V_{\beta_1}, V_{\beta_2}, \ldots, V_{\beta_m}$ covers C and C is compact.

Compactness may be characterized in terms of neighborhoods.

THEOREM 2.7 A topological space X is compact if and only if, whenever for each $x \in X$ a neighborhood N_x of x is given, there is a finite number of points $x_1, x_2, \ldots, x_n$ of X such that $X = \bigcup_{i=1}^{n} N_{x_i}$.

Proof. Suppose X is compact. Let there be given for each $x \in X$ a neighborhood N_x of x. For each x, there is an open set U_x such that $x \in U_x \subset N_x$ and consequently the family $\{U_x\}_{x \in X}$ is an open covering of X. Since X is compact there is a finite subcovering $U_{x_1}, U_{x_2}, \ldots, U_{x_n}$. But $U_{x_i} \subset N_{x_i}$ for each i, whence $N_{x_1}, N_{x_2}, \ldots, N_{x_n}$ covers X.

Conversely, suppose that whenever, for each $x \in X$ a neighborhood N_x of x is given, there is a finite number of points $x_1, x_2, \ldots, x_n$ of X such that $X = \bigcup_{i=1}^{n} N_{x_i}$. Let $\{U_\alpha\}_{\alpha \in I}$ be an open covering of X. Then, for each $x \in X$, there is an $\alpha = \alpha(x)$ such that $x \in U_\alpha$, and therefore $N_x = U_\alpha$ is a neighborhood of x. By our hypothesis, there are points $x_1, x_2, \ldots, x_n$ of X such that $N_{x_i} = U_{\alpha(x_i)}$, $i = 1, 2, \ldots, n$, covers X, and hence X is compact.

In terms of closed sets, we have:

THEOREM 2.8 A topological space is compact if and only if whenever a family $\{F_\alpha\}_{\alpha \in I}$ of closed sets is such that $\bigcap_{\alpha \in I} F_\alpha = \emptyset$ then there is a finite subset of indices $\{\alpha_1, \alpha_2, \ldots, \alpha_n\}$ such that $\bigcap_{i=1}^{n} F_{\alpha_i} = \emptyset$.

Proof. Suppose X is compact and a family $\{F_\alpha\}_{\alpha \in I}$ of closed sets is given such that $\bigcap_{\alpha \in I} F_\alpha = \emptyset$. Then

$$\bigcup_{\alpha \in I} C(F_\alpha) = C(\bigcap_{\alpha \in I} F_\alpha) = X,$$

so that $\{C(F_\alpha)\}_{\alpha \in I}$ is an open covering of X. Hence there is a finite subcovering $C(F_{\alpha_1}), C(F_{\alpha_2}), \ldots, C(F_{\alpha_n})$. Therefore

$$\bigcap_{i=1}^{n} F_{\alpha_i} = C\left(\bigcup_{i=1}^{n} C(F_{\alpha_i})\right) = \emptyset.$$

Conversely, suppose that for each family $\{F_\alpha\}_{\alpha \in I}$ of closed sets such that $\bigcap_{\alpha \in I} F_\alpha = \emptyset$ there is a finite subset of indices $\{\alpha_1, \alpha_2, \ldots, \alpha_n\}$ such that $\bigcap_{i=1}^{n} F_{\alpha_i} = \emptyset$. Let $\{O_\beta\}_{\beta \in J}$ be an open covering of X. Then $\{C(O_\beta)\}_{\beta \in J}$ is a family of closed sets such that $\bigcap_{\beta \in J} C(O_\beta) = \emptyset$. Thus $\bigcap_{i=1}^{n} C(O_{\alpha_i}) = \emptyset$ and $O_{\alpha_1}, O_{\alpha_2}, \ldots, O_{\alpha_n}$ is a finite subcovering.

THEOREM 2.9 Let $f: X \to Y$ be continuous and let A be a compact subset of X. Then $f(A)$ is a compact subset of Y.

Proof. Let $\{U_\alpha\}_{\alpha \in I}$ be an open covering of $f(A)$. Thus $f(A) \subset \bigcup_{\alpha \in I} U_\alpha$ and consequently

$$A \subset \bigcup_{\alpha \in I} f^{-1}(U_\alpha)$$

so that $\{f^{-1}(U_\alpha)\}_{\alpha \in I}$ is a covering of A. Since f is continuous, $f^{-1}(U_\alpha)$ is an open subset of X for each $\alpha \in I$ and therefore $\{f^{-1}(U_\alpha)\}_{\alpha \in I}$ is an open covering of A. A is compact, thus there is a finite subcovering $f^{-1}(U_{\alpha_1}), f^{-1}(U_{\alpha_2}), \ldots, f^{-1}(U_{\alpha_n})$ of A. But $A \subset f^{-1}(U_{\alpha_1}) \cup f^{-1}(U_{\alpha_2}) \cup \ldots \cup f^{-1}(U_{\alpha_n})$ implies that $f(A) \subset U_{\alpha_1} \cup U_{\alpha_2} \cup \ldots \cup U_{\alpha_n}$. $\{U_\alpha\}_{\alpha \in I}$ was an arbitrary open covering of $f(A)$, whence by Theorem 2.6, we have shown that $f(A)$ is compact.

COROLLARY 2.10 Let the topological spaces X and Y be homeomorphic. Then X is compact if and only if Y is compact.

Not every subset of a compact space is itself compact. We shall see that the closed interval $[0, 1]$ is compact, whereas the open interval $(0, 1)$ is not compact. To show that $(0, 1)$ is not compact, it suffices to find one open covering of $(0, 1)$ that does not have a finite subcovering. To this end, for each integer $n = 3, 4, 5, \ldots$, let $U_n = \left(\frac{1}{n}, 1 - \frac{1}{n}\right)$. Then $(U_n)_{n=3,4,5,\ldots}$ is an open covering of $(0, 1)$. On the other hand, for each integer $k > 3$ we have $\frac{1}{k} \notin \bigcup_{n=3}^{k} U_n$. Thus the union of every finite subcollection of $\{U_n\}_{n=3,4,5,\ldots}$ must fail to contain some point of $(0, 1)$, and hence there is no finite subcovering of $\{U_n\}_{n=3,4,5,\ldots}$.

We do, however, have this result.

THEOREM 2.11 Let X be compact. Then each closed subset of X is compact.

Proof. Let F be a closed subset of the compact space X. If $\{U_\alpha\}_{\alpha \in I}$ is an open covering of F, then by adjoining the open set $O = C(F)$ to the family $\{U_\alpha\}_{\alpha \in I}$ we obtain an open covering $\{V_\beta\}_{\beta \in J}$ of X. Since X is compact there is a finite subcovering $V_{\beta_1}, V_{\beta_2}, \ldots, V_{\beta_m}$ of X. But each V_{β_i} is either equal to a U_α for some $\alpha \in I$ or equal to O. If O occurs among $V_{\beta_1}, V_{\beta_2}, \ldots,$

V_{β_m} we may delete it to obtain a finite collection of the U_α's that covers $F = C(O)$.

Thus, in a compact space, for each subset the property of being closed implies the property of being compact. In a Hausdorff space, the converse is also true.

THEOREM 2.12 Let X be a Hausdorff space. If a subset F of X is compact, then F is closed.

Proof. We shall show that $O = C(F)$ is open by showing that for each point $z \in O$ there is a neighborhood U of z contained in O, or equivalently, for which $U \cap F = \emptyset$. To this end, with $z \in O$ fixed, by the Hausdorff property, we may choose for each point $x \in F$ an open neighborhood U_x of z and an open neighborhood V_x of x such that $U_x \cap V_x = \emptyset$. The family $\{V_x\}_{x \in F}$ is an open covering of F. Since F is compact, there is a finite subcovering $V_{x_1}, V_{x_2}, \ldots, V_{x_n}$ of F. The intersection $U = U_{x_1} \cap U_{x_2} \cap \ldots \cap U_{x_n}$ is an intersection of a finite set of neighborhoods of z and is therefore a neighborhood of z. Furthermore, U cannot intersect F since it does not intersect each element $V_{x_1}, V_{x_2}, \ldots, V_{x_n}$ of a covering of F. Thus $U \subset O$, from which it follows that O is a neighborhood of each of its points and $F = C(O)$ is closed.

COROLLARY 2.13 Let X be a compact Hausdorff space. Then a subset F of X is compact if and only if it is closed.

THEOREM 2.14 Let $f: X \to Y$ be a one-one continuous mapping of the compact space X onto a Hausdorff space Y. Then f is a homeomorphism.

Proof. We define $g: Y \to X$ by setting $g(y) = x$ if $f(x) = y$, so that f and g are inverse functions. It remains to prove that g is continuous. We shall prove this by proving that for each closed subset F of X, $g^{-1}(F)$ is a closed subset of Y. Given a closed subset F of X, by Theorem 2.11, F is compact. Hence

$f(F) = g^{-1}(F)$ is a compact subset of Y. By Theorem 2.12, $g^{-1}(F)$ is a closed subset of Y. Thus, g is continuous and f is a homeomorphism.

EXERCISES

1. Prove that the real line R is not compact.
2. Prove that every finite subset of a topological space is compact.
3. Let $\{U_\alpha\}_{\alpha \in I}$ be an open covering of $[0, 1]$. Define a subset P of $[0, 1]$ as follows: a point x is in P if and only if there is a finite collection $U_{\alpha_1}, U_{\alpha_2}, \ldots, U_{\alpha_m}$ of elements of $\{U_\alpha\}_{\alpha \in I}$ that covers $[0, x]$. Prove that if $x \in P$, then there is a neighborhood O of x such that $O \subset P$ and that therefore P is open. Prove that if $x \notin P$, then there is a neighborhood O of x such that $O \cap P = \emptyset$ and therefore P is closed. Conclude that $P = [0, 1]$ and that therefore $[0, 1]$ is compact.
4. Let X be a topological space. A family $\{F_\alpha\}_{\alpha \in I}$ of subsets of X is said to have the *finite intersection property* if for each finite subset J of I, $\bigcap_{\alpha \in J} F_\alpha \neq \emptyset$. Prove that X is compact if and only if for each family $\{F_\alpha\}_{\alpha \in I}$ of closed subsets of X that has the finite intersection property, we have $\bigcap_{\alpha \in I} F_\alpha \neq \emptyset$.
5. Let X be a set and $\mathfrak{J}$ and $\mathfrak{J}'$ be two topologies on X. Prove that if $\mathfrak{J} \subset \mathfrak{J}'$ and $(X, \mathfrak{J}')$ is compact then $(X. \mathfrak{J})$ is compact. Prove that if $(X, \mathfrak{J})$ is Hausdorff and $(X, \mathfrak{J}')$ is compact with $\mathfrak{J} \subset \mathfrak{J}'$, then $\mathfrak{J} = \mathfrak{J}'$.
6. Let $f: X \to Y$ be a continuous mapping of a compact space X onto a Hausdorff space Y. Prove that f is an identification.

3 COMPACT SUBSETS OF THE REAL LINE

DEFINITION 3.1 A subset A of R^n is said to be *bounded* if there is a real number K such that for each $x = (x_1, x_2, \ldots, x_n) \in A$, $|x_i| \leqq K$ for $1 \leqq i \leqq n$.

In particular a subset A of the real line R is bounded if A is contained in some closed interval $[-K, K]$, $K > 0$. Every closed interval $[a, b]$ is bounded for $[a, b] \subset [-K, K]$ where $K = \text{maximum } \{|a|, |b|\}$.

Lemma 3.2 If A is a compact subset of R then A is closed and bounded.

Proof. Since the real line satisfies the Hausdorff axiom, by Theorem 2.12, A is closed. For each positive integer n, let $O_n = (-n, n)$. $R = \bigcup_{n\in N} O_n$, where N is the set of natural numbers. Therefore $\{O_n\}_{n\in N}$ is an open covering of A. Since A is compact, $A \subset O_{n_1} \cup O_{n_2} \cup \ldots \cup O_{n_q}$. If we set k = maximum $\{n_1, n_2, \ldots, n_q\}$ then $O_{n_i} \subset O_k$ for $i = 1, 2, \ldots, q$, whence $A \subset O_k = (-k, k)$. Thus $A \subset [-k, k]$ and A is bounded.

Lemma 3.3 The closed interval $[0, 1]$ is compact.

Proof. Let $\{O_\alpha\}_{\alpha\in I}$ be a covering of $[0, 1]$ by open sets. Assume that there is no finite subcovering of $\{O_\alpha\}_{\alpha\in I}$. In this event, at least one of the two closed intervals $[0, \frac{1}{2}]$ or $[\frac{1}{2}, 1]$ cannot be covered by a finite subcollection of the family $\{O_\alpha\}_{\alpha\in I}$. Let $[a_1, b_1]$ denote one of these two intervals of length $\frac{1}{2}$ that cannot be covered by a finite subcollection of the family $\{O_\alpha\}_{\alpha\in I}$. We may now divide the interval $[a_1, b_1]$ into the two subintervals

$$\left[a_1, \frac{a_1 + b_1}{2}\right]$$

and

$$\left[\frac{a_1 + b_1}{2}, b_1\right]$$

of length $\frac{1}{4}$ and assert that at least one of these two intervals cannot be covered by a finite subcollection of the family $\{O_\alpha\}_{\alpha\in I}$. Let $[a_2, b_2]$ denote one of these two intervals of length $\frac{1}{4}$ that has the property that it cannot be covered by a finite subcollection of the family $\{O_\alpha\}_{\alpha\in I}$. We shall now proceed to define a sequence $[a_0, b_0]$, $[a_1, b_1]$, $[a_2, b_2], \ldots, [a_n, b_n], \ldots$ of such intervals. Assume that for $r = 0, 1, 2, \ldots, n$ we have defined intervals $[a_r, b_r]$ such that:

1. $[a_0, b_0] = [0, 1]$;

2. $b_r - a_r = \dfrac{1}{2^r}$ for $r = 0, 1, \ldots, n$;

3. for $r = 0, 1, \ldots, n-1$, either $[a_{r+1}, b_{r+1}] = \left[a_r, \frac{a_r + b_r}{2}\right]$ or $[a_{r+1}, b_{r+1}] = \left[\frac{a_r + b_r}{2}, b_r\right]$;
4. for each $r = 0, 1, \ldots, n$, no finite subcollection of $\{O_\alpha\}_{\alpha \in I}$ covers $[a_r, b_r]$.

We then define $[a_{n+1}, b_{n+1}]$ in the following manner. In view of (4) at least one of the two intervals

$$\left[a_n, \frac{a_n + b_n}{2}\right], \left[\frac{a_n + b_n}{2}, b_n\right]$$

cannot be covered by a finite subcollection of $\{O_\alpha\}_{\alpha \in I}$. Denote by $[a_{n+1}, b_{n+1}]$ whichever of these two intervals cannot be covered by a finite subcollection of $\{O_\alpha\}_{\alpha \in I}$, (if neither can be, we may agree that $[a_{n+1}, b_{n+1}]$ is the first of the two). Then conditions (2), (3), and (4) will also hold for $[a_{n+1}, b_{n+1}]$. It follows by induction that we may define a sequence $[a_0, b_0], [a_1, b_1], [a_2, b_2], \ldots$ of such intervals for which conditions (1) through (4) are true.

By conditions (3), $a_n \leqq a_{n+1} \leqq b_{n+1} \leqq b_n$. It follows that for each pair of positive integers m and n, $a_m \leqq b_n$. Thus each b_n is an upper bound of the set $\{a_0, a_1, a_2, \ldots\}$. Let a be the least upper bound of this set. Then $a \leqq b_n$ for each n, and hence a is a lower bound of the set $\{b_0, b_1, b_2, \ldots\}$. Let b be the greatest lower bound of the latter set. We therefore have $a \leqq b$. But, by the definition of a and b, we must have $a_n \leqq a \leqq b \leqq b_n$ for each n, whence by condition (2), $b - a \leqq \frac{1}{2^n}$ for each n and we conclude that $a = b$. We are now in a position to obtain a contradiction to condition (4), from which it will follow that our assumption that there is no finite subcovering of $[0, 1]$ is untenable.

$\{O_\alpha\}_{\alpha \in I}$ covers $[0, 1]$ and $a = b \in [0, 1]$. Therefore $a \in O_\beta$ for some $\beta \in I$. Since O_β is open there is an $\varepsilon > 0$ such that $B(a; \varepsilon) \subset O_\beta$. Let us choose the positive integer N large enough so that $\frac{1}{2^N} < \varepsilon$. Then $b_N - a_N < \varepsilon$. Now $a = b \in [a_N, b_N]$. Therefore, $a - a_N < \frac{1}{2^N} < \varepsilon$ and

$b - b_N < \frac{1}{2^N} < \varepsilon$. Consequently, $[a_N, b_N] \subset B(a; \varepsilon) \subset O_\beta$. Thus $[a_N, b_N]$ may be covered by a finite subcollection (namely, one!) of the family $\{O_\alpha\}_{\alpha \in I}$. Therefore the assumption that no finite subcollection of $\{O_\alpha\}_{\alpha \in I}$ covers $[0, 1]$ leads to a contradiction and we must conclude that $[0, 1]$ is compact.

It can be seen that the gist of the above argument is that if no finite subcollection of $\{O_\alpha\}_{\alpha \in I}$ covers $[0, 1]$, then no finite subcollection of $\{O_\alpha\}_{\alpha \in I}$ covers a sequence of subintervals whose lengths approach zero, whereas on the other hand if the length of one of these subintervals is small enough it is contained in some O_β.

Since each closed interval $[a, b]$ is homeomorphic to the closed interval $[0, 1]$ and compactness is a topological property, we obtain:

Corollary 3.4 Each closed interval $[a, b]$ is compact.

The next theorem, which characterizes the compact subsets of the real line, is frequently referred to as the Heine-Borel Theorem.

Theorem 3.5 A subset A of the real line is compact if and only if A is closed and bounded.

Proof. The "if" half of the theorem is Lemma 3.2. Conversely, if A is closed and bounded A is a closed subset of a closed interval $[-K, K]$ for some $K > 0$. But $[-K, K]$ is a compact space and therefore, by Theorem 2.11, A is compact.

EXERCISES

1. Using the method of subdivision of Lemma 3.3, prove that the unit square I^2 is a compact subset of the plane and in general that the unit n-cube I^n is a compact subspace of R^n.

2. Let X be a compact space and $(F_n)_{n=1,2,3,\ldots}$ a sequence of non-empty closed subsets of X such that $F_{n+1} \subset F_n$ for each n. Prove that $\bigcap_{n=1}^{\infty} F_n \neq \emptyset$.
3. Let $f: [a, b] \to R$ be continuous. Prove that the set $f([a, b])$ has both a least upper bound M and a greatest lower bound m and that there are points $u, v \in [a, b]$ such that $f(u) = M$, $f(v) = m$.
4. Let $f: [a, b] \to [c, d]$ be a continuous increasing function such that $f(a) = c$, $f(b) = d$. Prove that f is a homeomorphism.

4 PRODUCTS OF COMPACT SPACES

The fundamental result of this section is that the product of two compact spaces is itself compact. We shall establish this fact with the aid of the next lemma, which relates compactness to coverings by members of a base for the open sets. Let us recall that a base for the open sets of a topological space Z is a collection $\mathcal{B}$ of open subsets with the property that each open subset of Z is a union of members of the collection $\mathcal{B}$.

Lemma 4.1 Let $\mathcal{B}$ be a base for the open sets of a topological space Z. If, for each covering $\{B_\beta\}_{\beta \in J}$ of Z by members of $\mathcal{B}$, there is a finite subcovering, then Z is compact.

Proof. We must show that, if each covering of Z by "basic" open sets has a finite subcovering, then each open covering $\{O_\alpha\}_{\alpha \in I}$ of Z has a finite subcovering. For each $\alpha \in I$, O_α is a union of members of $\mathcal{B}$. Let J be an indexing set for all the basic sets B_β that occur in the expression of some O_α as a union of members of $\mathcal{B}$. Thus $\bigcup_{\alpha \in I} O_\alpha \subset \bigcup_{\beta \in J} B_\beta$ and hence $\{B_\beta\}_{\beta \in J}$ is a covering of Z by members of $\mathcal{B}$. It follows from our hypothesis that there is a finite subcovering $B_{\gamma_1}, B_{\gamma_2}, \ldots, B_{\gamma_n}$ of Z. Since each B_{γ_i} occurs in the expression of some O_{α_i}, $\alpha_i \in I$, as a union of members of $\mathcal{B}$, $B_{\gamma_i} \subset O_{\alpha_i}$. Consequently, $O_{\alpha_1}, O_{\alpha_2}, \ldots, O_{\alpha_n}$ must cover Z and therefore Z is compact.

Let us recall that if X and Y are topological spaces, then a base for open sets of $X \times Y$ is the collection of sets of the form $U \times V$, where U is open in X and V is open in Y.

THEOREM 4.2 Let X and Y be compact topological spaces; then $X \times Y$ is compact.

Proof. By virtue of Lemma 4.1 it suffices to prove that each covering of $X \times Y$ by sets of the form $U \times V$, U open in X, V open in Y, has a finite subcovering. Let $\{U_\alpha \times V_\alpha\}_{\alpha \in I}$ be such a covering. As an aid to understanding the proof, let us view the product $X \times Y$ as pictured in Figure 32, where a point $(x, y) \in X \times Y$ lies over the point $x \in X$ and level with the point $y \in Y$. In particular, for each $x_0 \in X$, the subset Y_{x_0} of $X \times Y$ consisting of all points (x_0, y), $y \in Y$, may be thought of as the collection of points lying over x_0. The open covering $\{U_\alpha \times V_\alpha\}_{\alpha \in I}$ is necessarily an open covering of Y_{x_0}. But Y_{x_0} is homeomorphic to Y and hence compact. We may therefore find a finite subset I_{x_0} of I such that $\{U_\alpha \times V_\alpha\}_{\alpha \in I_{x_0}}$ covers Y_{x_0} [this finite covering of Y_{x_0} is portrayed by the small rectangles in Figure 32]. We may also assume that $x_0 \in U_\beta$ for each $\beta \in I_{x_0}$, for

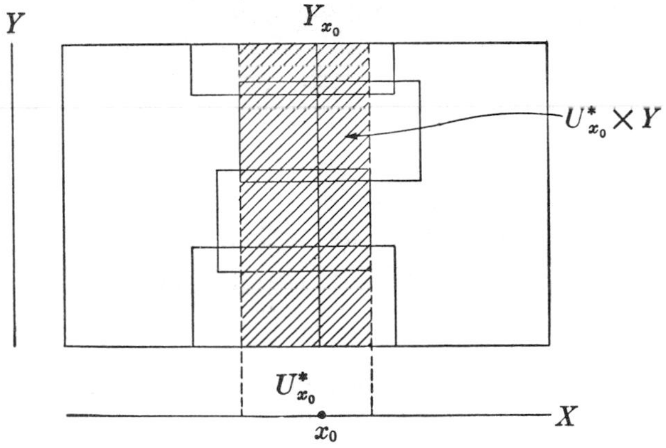

Figure 32

otherwise we may delete $U_\beta \times V_\beta$ and still cover Y_{x_0}. The set $U^*_{x_0} = \bigcap_{\alpha \in I_{x_0}} U_\alpha$ is a finite intersection of open sets containing x_0 and is therefore an open set containing x_0. We assert that $\{U_\alpha \times V_\alpha\}_{\alpha \in I_{x_0}}$ is an open covering of $U^*_{x_0} \times Y$. For, suppose $(x, y) \in U^*_{x_0} \times Y$. The point (x_0, y) is in $U_\beta \times V_\beta$ for some $\beta \in I_{x_0}$. Since $x \in U^*_{x_0}$, $x \in U_\alpha$ for all $\alpha \in I_{x_0}$. It follows that $(x, y) \in U_\beta \times V_\beta$, $\beta \in I_{x_0}$, proving our assertion.

Now $\{U^*_x\}_{x \in X}$ is an open covering of the compact space X, hence there is a finite subcovering $U^*_{x_1}, U^*_{x_2}, \ldots, U^*_{x_n}$ of X. Let us set $I^* = I_{x_1} \cup I_{x_2} \cup \ldots \cup I_{x_n}$ and show that the finite family $\{U_\alpha \times V_\alpha\}_{\alpha \in I^*}$ is a covering of $X \times Y$. Given a point $(x, y) \in X \times Y$, $x \in U^*_{x_i}$ for some x_i so that $(x, y) \in U^*_{x_i} \times Y$. By our previous assertion $(x, y) \in U_\beta \times V_\beta$ for some $\beta \in I_{x_i}$ which certainly implies that $(x, y) \in U_\alpha \times V_\alpha$ for some $\alpha \in I^*$. We have thus established that $\{U_\alpha \times V_\alpha\}_{\alpha \in I^*}$ is a finite subcovering and that therefore $X \times Y$ is compact.

If $X_1, X_2, \ldots, X_n$ are topological spaces, one may distinguish between $\prod_{i=1}^{n} X_i$ and $\left(\prod_{i=1}^{n-1} X_i\right) \times X_n$, for the points of the first space are n-tuples $(x_1, x_2, \ldots, x_n)$, whereas the points of the second space are ordered pairs $((x_1, x_2, \ldots, x_{n-1}), x_n)$ whose first elements are $(n - 1)$-tuples. Nevertheless, these two spaces are certainly homeomorphic [the obvious homeomorphism carries a point $(x_1, x_2, \ldots, x_n)$ into $((x_1, x_2, \ldots, x_{n-1}), x_n)$], hence by induction on n we obtain:

Corollary 4.3 Let $X_1, X_2, \ldots, X_n$ be compact topological spaces. Then $\prod_{i=1}^{n} X_i$ is also compact.

It is true that the product of an arbitrary family of compact spaces is compact. This result, which we shall not prove, is called the Tychonoff Theorem.

Let us recall that the unit n-cube I^n is the subset of R^n consisting of all points $x = (x_1, x_2, \ldots, x_n)$ such that $0 \leqq x_i \leqq 1$ for $i = 1, 2, \ldots, n$. As a subspace of R^n, I^n has the same topology as the product space $I \times I \times \ldots \times I$ (n-factors). Since $I = [0, 1]$ is compact, as a special case of Corollary 4.3 we have:

COROLLARY 4.4 The unit n-cube I^n is compact.

THEOREM 4.5 A subset A of R^n is compact if and only if A is closed and bounded.

Proof. The proof that if A is compact then A is closed and bounded is similar to the proof of this fact for a subset of the real line as presented in the preceding section. Conversely, we shall first show that each closed "cube" is compact. The collection of points $x = (x_1, x_2, \ldots, x_n)$ in R^n such that $|x_i| \leqq K$ for $i = 1, 2, \ldots, n$, which we shall denote by M_K, is a "cube" of width $2K$ with center at the origin. M_K is homeomorphic to the unit n-cube I^n, for the function $F: I^n \to M_K$ defined by

$$F(x_1, x_2, \ldots, x_n) = (2Kx_1 - K, 2Kx_2 - K, \ldots, 2Kx_n - K)$$

is easily seen to be a homeomorphism (Theorem 2.14). Since I^n is compact, M_K is compact. Now suppose A is closed and bounded; then A is a closed subset of the compact cube M_K for some K, whence A is compact.

EXERCISES

1. Let S be the set $[0, 1]$ and define a subset F of S to be closed if either it is finite or is equal to S. Prove that this definition of closed set yields a topology for S. Show that S with this topology is connected, path-wise connected, and compact, but that S is not a Hausdorff space. Show that each subset of S is compact and that therefore there are compact subsets of S that are not closed.
2. A topological space X is said to be *locally compact* if each point $x \in X$ has at least one compact neighborhood. Prove that the real line and R^n are locally compact.

3. Let X be a topological space and x^* a point of X. Assume a base for the system of neighborhoods of x^* consists of the complements of compact subsets of $X - \{x^*\}$. Prove X is compact. Prove that if in addition $X - \{x^*\}$ is a locally compact Hausdorff space, then X is a compact Hausdorff space. Given a locally compact Hausdorff space Y which is not compact, show that Y is a subspace of a compact Hausdorff space that contains one more point than Y does. This space is called the *one-point compactification* of Y. Prove that the one-point compactification of R^n is homeomorphic to S^n.

5 COMPACT METRIC SPACES

A metric space (X, d) is said to be *compact* or is called a *compactum* if its associated topological space is compact. In this section we shall derive certain properties of compact metric spaces. A basic result is that a metric space is compact if and only if every infinite subset has at least one "point of accumulation."

DEFINITION 5.1 Let X be a topological space and A a subset of X. A point $a \in X$ is called an *accumulation point* of A if each neighborhood of a contains infinitely many distinct points of A.

In referring to the accumulation points of a set A, care must be taken to specify of which topological space A is to be considered a subset. For example, in the real line R, the subset $A = \left\{1, \frac{1}{2}, \frac{1}{3}, \ldots, \frac{1}{n}, \ldots\right\}$ has the accumulation point 0, whereas in the topological space $(0, +\infty)$, the same set A has no accumulation points.

Recall that in a metric space we defined a as a limit point of a subset A if every neighborhood of a contains a point of A different from a. If we use the same definition in a topological space every accumulation point of A is also a limit point of A.

In Hausdorff spaces, and hence in metric spaces, accumulation points and limit points coincide.

LEMMA 5.2 Let X be a Hausdorff space and A a subset of X. A point $a \in X$ is an accumulation point of A if and only if a is a limit point of A.

Proof. Suppose a is not an accumulation point of A. Then there is a neighborhood N of a that contains at most a finite collection $\{a_1, a_2, \ldots, a_p\}$ of points of A distinct from a. For each of these points a_i, $i = 1, 2, \ldots, p$, we can find neighborhoods U_i of a and neighborhoods V_i of a_i such that $U_i \cap V_i = \emptyset$. Then $N \cap U_1 \cap U_2 \cap \ldots \cap U_p$ is a neighborhood of a that contains no points of A other than possibly a.

THEOREM 5.3 Let X be a compact space; then every infinite subset K of X has at least one limit point in X.

Proof. Suppose K is a subset of X that has no limit points. For each $x \in K$ there is a neighborhood N_x of x such that $N_x \cap K = \{x\}$. K is closed and hence compact. Therefore there are points $x_1, x_2, \ldots, x_m$ such that $N_{x_1}, N_{x_2}, \ldots, N_{x_m}$ cover K. It follows that $K = \{x_1, x_2, \ldots, x_m\}$ and K is finite.

For compact Hausdorff spaces, and in particular for compact metric spaces, Theorem 5.3 becomes every infinite subset A of X has at least one point of accumulation in X. The next two lemmas are used to prove the converse for metric spaces.

LEMMA 5.4 Let (X, d) be a metric space such that every infinite subset of X has at least one accumulation point in X. Then, for each positive integer n, there is a finite set of points $x_1^n, x_2^n, \ldots, x_p^n$ of X such that the collection of open balls

$$B\left(x_1^n; \frac{1}{n}\right), B\left(x_2^n; \frac{1}{n}\right), \ldots, B\left(x_p^n; \frac{1}{n}\right)$$

covers X.

Proof. Suppose there were an integer n such that no finite collection of balls of radius $\frac{1}{n}$ covered X. Choose a point $x_1 \in X$. $B\left(x_1; \frac{1}{n}\right)$ certainly does not cover X, hence there is a point $x_2 \in X$ such that $x_2 \notin B\left(x_1; \frac{1}{n}\right)$. $B\left(x_1; \frac{1}{n}\right) \cup B\left(x_2; \frac{1}{n}\right)$ does not cover X, hence there is a point $x_3 \in X$ such that $x_3 \notin B\left(x_1; \frac{1}{n}\right) \cup B\left(x_2; \frac{1}{n}\right)$. Continuing in this way we may construct a sequence $x_1, x_2, \ldots, x_k, \ldots$ of points of X such that for $k > 1$,

$$x_k \notin \bigcup_{i=1}^{k-1} B\left(x_i; \frac{1}{n}\right).$$

Thus

$$d(x_k, x_{k'}) \geqq \frac{1}{n}$$

if $k \neq k'$. It follows that the set $\{x_1, x_2, \ldots, x_k, \ldots\}$ is infinite and therefore has a point of accumulation $x \in X$. The neighborhood $B\left(x; \frac{1}{2n}\right)$ contains infinitely many points of $\{x_1, x_2, \ldots, x_k, \ldots\}$ and in particular contains two points $x_k, x_{k'}$ with $k \neq k'$. Since $x_k, x_{k'} \in B\left(x; \frac{1}{2n}\right)$, we obtain the contradiction $d(x_k, x_{k'}) < \frac{1}{n}$.

A similar argument yields the following result.

Lemma 5.5 Let (X, d) be a metric space such that each infinite subset of X has at least one point of accumulation. Then for each open covering $\{O_\alpha\}_{\alpha \in I}$ of X there is a positive number ε such that each open ball $B(x; \varepsilon)$ is contained in an element O_β of this covering.

Proof. We shall suppose the result is false and obtain a contradiction. If the lemma is false, for each

$n = 1, 2, \ldots$, there is an open ball $B\left(x_n; \frac{1}{n}\right)$ such that $B\left(x_n; \frac{1}{n}\right) \not\subset O_\alpha$ for each $\alpha \in I$. Let $A = \{x_1, x_2, \ldots\}$. If A is finite, some point $x \in X$ occurs infinitely often in the sequence $x_1, x_2, \ldots$. Since $\{O_\alpha\}_{\alpha \in I}$ covers X, $x \in O_\beta$ for some $\beta \in I$. Also, O_β is open, hence there is a $\delta > 0$ such that $B(x; \delta) \subset O_\beta$. We may, however, choose n so that $\frac{1}{n} < \delta$ and $x_n = x$, in which case

$$B\left(x_n; \frac{1}{n}\right) = B\left(x; \frac{1}{n}\right) \subset O_\beta,$$

a contradiction. There remains the possibility that $A = \{x_1, x_2, \ldots\}$ is infinite. Thus A has at least one point of accumulation x. Again $x \in O_\beta$ for some $\beta \in I$ so that $B(x; \delta) \subset O_\beta$ for some $\delta > 0$. There are infinitely many points of A in the neighborhood $B\left(x; \frac{\delta}{2}\right)$ of x. Hence we may choose an n such that $\frac{1}{n} < \frac{\delta}{2}$ and $x_n \in B\left(x; \frac{\delta}{2}\right)$. We then have $B\left(x_n; \frac{1}{n}\right) \subset B(x; \delta) \subset O_\beta$, which is again a contradiction.

The number ε of Lemma 5.5 depends on the particular open covering $\{O_\alpha\}_{\alpha \in I}$ considered. Given the open covering $\{O_\alpha\}_{\alpha \subset I}$, if the number ε has the property that for each $x \in X$, $B(x; \varepsilon) \subset O_\beta$ for some $\beta \in I$, then each number ε' with $0 < \varepsilon' < \varepsilon$ also has this property. The least upper bound of the set of numbers having this property is called the *Lebesgue number*, ε_L, of the open covering $\{O_\alpha\}_{\alpha \in I}$. We may now state:

Corollary 5.6 Let (X, d) be a metric space such that each infinite subset of X has an accumulation point. Then each open covering $\{O_\alpha\}_{\alpha \in I}$ of X has a Lebesgue number ε_L.

A topological space X is said to have the *Bolzano-Weierstrass property* if each infinite subset of X has at least one point of accumulation. We may now prove that every metric space that has the Bolzano-Weierstrass property is a compact metric space.

THEOREM 5.7 Let (X, d) be a metric space that has the property that every infinite subset of X has at least one accumulation point. Then X is compact.

Proof. Let $\{O_\alpha\}_{\alpha \in I}$ be an open covering and let ε_L be its Lebesque number. Let us choose n so that $\frac{1}{n} < \varepsilon_L$. By Lemma 5.4 there is a finite set $\{x_1, x_2, \ldots, x_p\}$ of points of X such that the open balls $B\left(x_1; \frac{1}{n}\right)$, $B\left(x_2; \frac{1}{n}\right), \ldots, B\left(x_p; \frac{1}{n}\right)$ cover X. Furthermore, by Lemma 5.5, for each $i = 1, 2, \ldots, p$, there is a $\beta_i \in I$ such that $B\left(x_i; \frac{1}{n}\right) \subset O_{\beta_i}$. It follows that the collection $O_{\beta_1}, O_{\beta_2}, \ldots, O_{\beta_p}$ is a finite subcovering of $\{O_\alpha\}_{\alpha \in I}$.

We have now proved the main result of this section.

THEOREM 5.8 Let (X, d) be a metric space. Each infinite subset of X has at least one accumulation point if and only if X is compact.

Having proved that a subspace X of Euclidean n-space R^n is compact if and only if it is closed and bounded, we may state:

COROLLARY 5.9 Let X be a subspace of R^n. Then the following three properties are equivalent:

1. X is compact.
2. X is closed and bounded.
3. Each infinite subset of X has at least one point of accumulation in X.

The existence, for each open covering of a compact metric space, of a Lebesgue number has as a consequence the fact that each continuous function defined on a compact metric space is "uniformly" continuous.

DEFINITION 5.10 Let $f:(X, d) \to (Y, d')$ be a function from a metric space (X, d) to a metric space (Y, d'). f is said to be *uniformly continuous* if, for each positive number ε, there is a $\delta > 0$, such that whenever $d(x, y) < \delta$, then $d'(f(x), f(y)) < \varepsilon$.

If the function $g:X \to Y$ is continuous, then for each $x \in X$ and each $\varepsilon > 0$, there is a $\delta > 0$, where δ may depend on both the choice of x and ε, such that $d(x, a) < \delta$ implies $d'(g(x), g(a)) < \varepsilon$. If, however, g is uniformly continuous, then given $\varepsilon > 0$, the number δ may be used at each point $x \in X$, that is, uniformly throughout X, to yield $d'(g(x), g(a)) < \varepsilon$ if $d(x, a) < \delta$. Thus:

COROLLARY 5.11 If $f:X \to Y$ is uniformly continuous, then f is continuous.

On the other hand a continuous function need not be uniformly continuous. As an example, consider $f:(0, 1] \to R$ defined by $f(x) = \frac{1}{x}$. Given $\varepsilon = 1$, we shall show that there does not exist a $\delta > 0$ such that $|x - y| < \delta$ implies $|f(x) - f(y)| < 1$. For given any $\delta > 0$ we can choose n large enough so that if $x = \frac{1}{n}$, $y = \frac{1}{n+1}$ we have

$$x - y = \frac{1}{n(n+1)} < \delta$$

whereas

$$\left|f\left(\frac{1}{n}\right) - f\left(\frac{1}{n+1}\right)\right| = 1.$$

In view of the next result, it should be noted that in this example the interval $(0, 1]$ is not compact.

THEOREM 5.12 Let $f: (X, d) \to (Y, d')$ be a continuous function from a compact metric space X to a metric space Y. Then f is uniformly continuous.

Proof. Given $\varepsilon > 0$, for each $x \in X$, there is a $\delta_x > 0$ such that if $y \in B(x; \delta_x)$ then $f(y) \in B\left(f(x); \frac{\varepsilon}{2}\right)$. The collection $\{B(x; \delta_x)\}_{x \in X}$ is an open covering of X. Since X is compact, this open covering has a Lebesgue number. Let us choose δ to be a positive number less than this Lebesgue number. If $z, z' \in X$ and $d(z, z') < \delta$ so that z and z' are in a ball of radius less than δ, we have $z, z' \in B(x; \delta_x)$ for some $x \in X$. Consequently, $f(z), f(z') \in B\left(f(x); \frac{\varepsilon}{2}\right)$, whence $d'(f(z), f(z')) \leqq d'(f(z), f(x)) + d'(f(x), f(z')) < \varepsilon$.

EXERCISES

1. In a metric space (X, d), a sequence $a_1, a_2, \ldots$ of points of X is called a *Cauchy sequence* if for each $\varepsilon > 0$ there is a positive integer N such that $d(a_n, a_m) < \varepsilon$ whenever $n, m > N$. A metric space X is called *complete* if every Cauchy sequence in X converges to a point of X. Prove that a compact metric space is complete.
2. In Euclidean n-space R^n, prove that every Cauchy sequence lies in a bounded closed subset of R^n. Use this fact to prove that R^n is complete.
3. Let (X, d) be a compact metric space. Prove that X is "bounded with respect to d"; that is, there is a positive number K such that $d(x, y) \leqq K$ for all $x, y \in X$.
4. Let (X, d) be a compact metric space and let $\{F_\alpha\}_{\alpha \in I}$ be a family of closed subsets of X such that $\bigcap_{\alpha \in I} F_\alpha = \emptyset$. Prove that there is a positive number c such that for each $x \in X$, $d(x, F_\alpha) \geqq c$ for some $\alpha \in I$.
5. A subset A of a topological space X is called *dense* if $\overline{A} = X$. Let X be a compact metric space. Prove that there is a sequence $a_1, a_2, \ldots$ of points of X such that the set $A = \{a_1, a_2, \ldots\}$ is dense in X.

6. Let X be the set of continuous functions $f:[a, b] \to R$. Let $I: X \to R$ be defined by $I(f) = \int_a^b f(t)\,dt$. Define a distance function d on X by setting $d(f, g) = \underset{a \leqq t \leqq b}{\text{l.u.b.}} |f(t) - g(t)|$. Prove that I is uniformly continuous. Let $f_1, f_2, \ldots$ be a Cauchy sequence in (X, d). Prove that for each $t \in [a, b]$, $f_1(t), f_2(t), \ldots$ is a Cauchy sequence of real numbers. For each $t \in [a, b]$, denote by $f(t)$ the limit of this sequence. Prove that the function $f:[a, b] \to R$ so defined is continuous, that $\lim_n f_n = f$ in X, and therefore (X, d) is complete, so that in the terminology of Problems 2 and 3 in Section 8, Chapter 2, X is a complete normed vector space. [A complete normed vector space with either the real or complex numbers as scalars is called a *Banach space.*]
7. Let A be any set and let R^A be the set of all functions $f: A \to R$ where R is the reals. Define $f + g$ by $(f + g)(a) = f(a) + g(a)$ and αf by $(\alpha f)(a) = \alpha f(a)$, for $f, g \in R^A$ and $\alpha \in R$. Prove that R^A is a vector space with R as scalars. A function $f \in R^A$ is bounded if $||f|| = \underset{a \in A}{\text{l.u.b.}} |f(a)|$ exists. Prove that the set B of bounded functions is a normed vector space in the sense of Problem 2, Section 8, Chapter 2. Prove that B is a complete metric space. Now let A be a topological space and let $C(A, R)$ be the set of all bounded continuous functions from A to R. Prove that $C(A, R)$ is a closed subset of B and is complete.

6 COMPACTNESS AND THE BOLZANO-WEIERSTRASS PROPERTY

Theorem 5.8, which states that a metric space is compact if and only if each infinite subset has at least one accumulation point, raises the question as to whether or not these two properties are equivalent in an arbitrary topological space. We already know that the first implies the second for Hausdorff spaces. Since there are examples of topological spaces that are not compact, but in which each infinite subset has a point of accumulation, compactness is the stronger of the two properties. We might therefore think of the second property, which we have called the Bolzano-

Weierstrass property, as a weaker type of compactness. To illustrate how much weaker the Bolzano-Weierstrass property is, we need to introduce the concept of countability.

DEFINITION 6.1 A non-empty set X is said to be *countable* if there is an onto function $f: N \to X$, where N is the set of positive integers.

A finite set $X = \{x_1, x_2, \ldots, x_n\}$ is countable, for we may construct an onto function $f: N \to X$ by setting $f(i) = x_i$, $1 \leqq i \leqq n$, and defining $f(i)$ for $i > n$ arbitrarily, say $f(i) = x_1$, $i > n$. A countable set that is not finite is called *denumerable*. In this case an onto function $f: N \to X$ gives rise to an "enumeration," $x_1 = f(1)$, $x_2 = f(2)$, . . . , $x_n = f(n)$, . . . of the elements of X, so that we may write $X = \{x_1, x_2, \ldots, x_n, \ldots\}$. Since we have not required the function f to be one-one, a given element $x \in X$ may occur more than once in this enumeration. It is easy to see, however, that by deleting all but the first occurrence of any element $x \in X$ and reducing the succeeding subscripts accordingly, it is possible to obtain an enumeration of X in which each element occurs one and only one time.

There are several facts about countability that are of general interest. As a simple consequence of Definition 6.1 we obtain:

COROLLARY 6.2 Let X and Y be non-empty sets. If X is countable and there is an onto function $g: X \to Y$, then Y is countable.

Proof. Since X is countable, there is an onto function $f: N \to X$, N the set of positive integers. The composite function $gf: N \to Y$ is onto and hence Y is countable.

COROLLARY 6.3 A non-empty subset of a countable set is countable.

Proof. Let $A \subset X$, X countable, A non-empty. We may define an onto function $g: X \to A$ by setting $g(a) = a$ for $a \in A$ and defining g arbitrarily for points $x \notin A$.

The set N of positive integers is countable, since the identity function $i:N \to N$ is onto. On the other hand, the collection 2^N of subsets of N is not countable, since for an arbitrary set X there is no onto function $f:X \to 2^X$ [see Exercise 1]. A set that is not countable is called *uncountable.* Another example of an uncountable set is the set R of real numbers [see Exercise 2]. Surprisingly, $N \times N$ is a countable set.

THEOREM 6.4 Let N be the set of positive integers. Then $N \times N$ is countable.

Proof. The elements of $N \times N$ may be arrayed in the form of the infinite matrix of Figure 33. We may arrange these elements in the form of a sequence,

$$\begin{matrix} (1, 1) & (1, 2) & (1, 3) & \cdots & (1, n) & \cdots \\ (2, 1) & (2, 2) & (2, 3) & \cdots & (2, n) & \cdots \\ (3, 1) & (3, 2) & (3, 3) & \cdots & (3, n) & \cdots \\ \vdots & \vdots & \vdots & \vdots & \vdots & \vdots \\ (m, 1) & (m, 2) & (m, 3) & \cdots & (m, n) & \cdots \\ \vdots & \vdots & \vdots & \vdots & \vdots & \vdots \end{matrix}$$

Figure 33

$x_1 = f(1), x_2 = f(2), \ldots, x_k = f(k), \ldots,$ by setting $x_1 = (1, 1), x_2 = (2, 1), x_3 = (1, 2), x_4 = (3, 1), \ldots;$ that is, having exhausted the entries on the diagonal of this matrix from $(p, 1)$ to $(1, p)$ we proceed to enumerate the entries on the diagonal from $(p + 1, 1)$ to $(1, p + 1)$. To explicitly define the onto function $f:N \to N \times N$ we note that there are $\dfrac{p(p + 1)}{2}$ entries on or above the diagonal from $(p, 1)$ to $(1, p)$, hence if $1 \leqq j \leqq p + 1$ we are setting

$$x_{\frac{p^2+p}{2}+j} = f\left(\frac{p^2+p}{2} + j\right) = (p - j + 2, j).$$

As a direct consequence of Theorem 6.4 and Corollary 6.2 one obtains the result that the set Q^+ of positive rational numbers is countable, for the function $h:N \times N \to Q^+$ defined by $h(r, s) = \frac{r}{s}$, $(r, s) \in N \times N$ is onto.

COROLLARY 6.5 Let $X_1, X_2, \ldots, X_n, \ldots,$ be a sequence of sets, each of which is countable. Then $\bigcup_{i=1}^{\infty} X_i$ is a countable set.

Proof. Since each X_i is countable there is an onto function $f_i:N \to X_i$, $i = 1, 2, \ldots, n, \ldots.$ We define a function $F:N \times N \to \bigcup_{i=1}^{\infty} X_i$ by setting $F(i, j) = f_i(j)$, $(i, j) \in N \times N$. F is onto, for if $x \in \bigcup_{i=1}^{\infty} X_i$, $x \in X_i$ for some i, whence $x = f_i(j) = F(i, j)$ for some $(i, j) \in N \times N$. But $N \times N$ is countable and therefore $\bigcup_{i=1}^{\infty} X_i$ is countable.

A more direct proof of Corollary 6.5 can be given by utilizing the matrix of Figure 33 to display the elements of $\bigcup_{i=1}^{\infty} X_i$, entering the element $x_j^i = f_i(j) = F(i, j)$ in the ith row and j^{th} column. One then enumerates the elements of $\bigcup_{i=1}^{\infty} X_i$ in accordance with the scheme adopted in the proof of Theorem 6.4. Since any countable collection of sets may be arranged in the form of a finite or infinite sequence of sets, Corollary 6.5 states that, if X is the union of a countable collection of sets, each of which is countable, then X is countable.

In view of the fact that the set Q^+ of positive rational numbers is countable, the set Q^- of negative rational numbers is also countable. Consequently, the set Q of all rational numbers is countable. Using Corollary 6.5 we may then assert that the collection B of all open intervals on the real line of the form $B(p; q)$, $q > 0$, with p and q rational, is also a countable set, for it is a countable union of sets each of which is countable. This fact may be used to prove that there is a countable basis for the open sets on the real line.

Let us now return to our discussion of the relation between compactness and the Bolzano-Weierstrass property. The Bolzano-Weierstrass property implies that each countable covering has a finite subcovering.

THEOREM 6.6 Let E be a subspace of a topological space X with the property that each infinite subset of E has a point of accumulation in E. Then every countable open covering of E has a finite subcovering.

Proof. We may assume that a countable open covering of E is given in the form of a sequence $O_1, O_2, \ldots, O_n, \ldots$ of open subsets of X such that $E \subset \bigcup_{n=1}^{\infty} O_n$. Suppose that no finite subcollection covers E. Then for each integer k, the open set $O_k^* = \bigcup_{n=1}^{k} O_n$ does not cover E. Hence for each k there is a point $x_k \in E$ such that $x_k \notin O_k^*$. The subset

$$A = \{x_1, x_2, \ldots, x_k, \ldots\}$$

of E must be infinite. Let $x \in E$ be a point of accumulation of A. Since $x \in E$, $x \in O_p$ for some index p. O_p is a neighborhood of x and therefore infinitely many of the points of A belong to O_p. In particular, for some $k > p$ we would have $x_k \in O_p \subset O_p^* \subset O_k^*$, contradicting the choice of x_k. Therefore there must be a finite subcollection of the open sets $O_1, O_2, \ldots, O_n, \ldots$ that covers E.

If a topological space X is such that every open covering has a countable subcovering, by virtue of Theorem 6.6, the Bolzano-

Weierstrass property implies compactness. A sufficient condition for every open covering to have a countable subcovering is given by the next theorem, often called Lindelöf's Theorem.

THEOREM 6.7 Let X be a topological space that has a countable basis for the open sets. Then each open covering $\{O_\alpha\}_{\alpha \in I}$ has a countable subcovering.

Proof. Let $\mathcal{B} = \{B_\beta\}_{\beta \in J}$ be a countable basis for the open sets of X. We shall first prove that for each point $x \in X$ and each open set O containing x, there is a basis element B_β such that $x \in B_\beta \subset O$. For, since $\mathcal{B}$ is a basis for the open sets, O is a union of elements of $\mathcal{B}$, thus $O = \bigcup_{\beta \in J'} B_\beta$ for some subset J' of J. But $x \in O$, hence $x \in B_\beta$ for some $\beta \in J'$, and clearly $B_\beta \subset O$. Now suppose that $\{O_\alpha\}_{\alpha \in I}$ is an open covering of x. We must find a countable subset $I' \subset I$ such that $\{O_\alpha\}_{\alpha \in I'}$ is a covering. For each $x \in X$ and each O_α containing x, we choose a B_β such that $x \in B_\beta \subset O_\alpha$. The totality of sets B_β so chosen constitute a countable subfamily $\{B_\beta\}_{\beta \in J'}$ of the basis $\mathcal{B}$ and this subfamily covers X. Now, for each such B_β with $\beta \in J'$, let us choose a single index $\alpha = f(\beta) \in I$ such that $B_\beta \subset O_\alpha = O_{f(\beta)}$. The totality of sets O_α so chosen constitute a subfamily $\{O_\alpha\}_{\alpha \in I'} = \{O_{f(\beta)}\}_{\beta \in J'}$, which is also countable and must cover X, for $\bigcup_{\beta \in J'} B_\beta \subset \bigcup_{\alpha \in I'} O_\alpha$.

COROLLARY 6.8 Let X be a topological space that has a countable basis for the open sets. Then X is compact if and only if X has the Bolzano-Weierstrass property.

Although we shall not give an example of a topological space X that has the Bolzano-Weierstrass property, but is not compact, the preceding discussion has revealed that such a space must be found among those topological spaces which are not metrizable and do not possess a countable basis for the open sets. Those spaces which possess a countable base for the open sets are called *completely separable* or are said to satisfy the *second axiom of countability.*

EXERCISES

1. Let X be an arbitrary non-empty set and $f:X \to 2^X$ an arbitrary function from X to the subsets of X. Let A be the subset of X consisting of those points $x \in X$ such that $x \notin f(x)$. Prove that there cannot be a point $a \in X$ such that $A = f(a)$. Finally, prove that there is no onto function $f:X \to 2^X$.
2. Let a function $f:N \to [0, 1]$ be given, N the set of positive integers. In the resulting enumeration $x_1 = f(1), x_2 = f(2), \ldots$, of numbers in $[0, 1]$, express each number x_k in decimal notation $x_k = .a_1^k a_2^k \ldots a_n^k \ldots$, a_n^k an integer $0 \leq a_n^k \leq 9$. Construct a real number $y = .y_1 y_2 \ldots y_n \ldots$ such that $y_r \neq a_r^r$, $r = 1, 2, \ldots$, thereby obtaining the result that f cannot be onto and consequently the real numbers are not countable.
3. Use the rational density theorem, which states that between any two real numbers there is a rational number, to prove that the collection of open intervals $B(p; q)$, $q > 0$, p, q rational are a basis for the open sets of R and that therefore R satisfies the second axiom of countability.
4. Let X and Y be topological spaces satisfying the second axiom of countability. Prove that $X \times Y$ also satisfies the second axiom of countability and hence R^n does.
5. Let $\{A_\alpha\}_{\alpha \in I}$ and $\{B_\beta\}_{\beta \in J}$ be families of subsets of a set X. $\{A_\alpha\}_{\alpha \in I}$ is called a *refinement* of $\{B_\beta\}_{\beta \in J}$ if for each $\alpha \in I$ there is a $\beta \in J$ such that $A_\alpha \subset B_\beta$. Suppose that $\{A_\alpha\}_{\alpha \in I}$ is a refinement of $\{B_\beta\}_{\beta \in J}$ and that $\{A_\alpha\}_{\alpha \in I}$ covers X. Prove that if I is finite there is a finite subcovering of $\{B_\beta\}_{\beta \in J}$ and if I is countable there is a countable subcovering of $\{B_\beta\}_{\beta \in J}$.
6. Recall that a subset A of a topological space X is called *dense* in X if $\overline{A} = X$. A topological space X is called *separable* if there is a countable dense subset. Prove that X is separable if X satisfies the second axiom of countability.
7. A topological space X is said to satisfy the *first axiom of countability* if at each point $x \in X$ there is a countable basis for the complete system of neighborhoods at x. Prove that if X satisfies the second axiom of countability then X satisfies the first axiom of countability.
8. Let X satisfy the first axiom of countability. Prove that for each

$x \in X$ there is a countable basis $U_1, U_2, \ldots$ for the neighborhoods at x such that $U_1 \supset U_2 \supset \ldots$ and such that if $u_i \in U_i$ then $\lim_n u_n = x$. Let $f: X \to Y$ be a function into a second topological space Y. Show that if for all sequences $x_1, x_2, \ldots$ such that $x = \lim_n x_n$ we have $f(x) = \lim_n f(x_n)$ then f is continuous at x.

9. Let $f: X \to X$ be a function from a metric space X into itself. f is said to be *contractive* if there is a positive constant $K < 1$ such that $d(f(x), f(x')) < K \cdot d(x, x')$ for all $x, x' \in X$. Prove that a contractive function is continuous. Let $a \in X$. Set $a_0 = a$, $a_1 = f(a)$, $a_2 = f(a_1)$, and in general $a_{n+1} = f(a_n) = f^{n+1}(a)$. Prove that for such an f the following hold: $d(a_{n+1}, a_n) < K^{n-1} d(a_1, a_0)$; $a_0, a_1, \ldots, a_n, \ldots$ is a Cauchy sequence. If X is a complete metric space so that $\lim_n a_n = a$ for some $a \in X$, then a is a fixed point of f, and if $f(b) = b$, $b = a$, so that every contractive mapping has a unique fixed point.

7 SURFACES BY IDENTIFICATION

In an earlier section we discussed the function $p: [0, 1] \to S^1$ defined by $p(t) = (\cos 2\pi t, \sin 2\pi t)$. p is a continuous function defined on a compact space and onto a Hausdorff space. Whenever this is the case the topology of the image space is determined by the function and the domain space.

LEMMA 7.1 Let $f: X \to Y$ be a continuous mapping of a compact space X onto a Hausdorff space Y. Then a subset B of Y is closed if and only if $f^{-1}(B)$ is a closed subset of X.

Proof. This lemma is a weaker form of Theorem 2.14. First suppose B is closed. Then $f^{-1}(B)$ is closed by the continuity of f. Conversely, if $f^{-1}(B)$ is closed, then $f^{-1}(B)$ is compact. $B = f(f^{-1}(B))$, hence B is compact. Being a compact subset of a Hausdorff space, B is closed.

Corollary 7.2 Let $f:X \rightarrow Y$ be a continuous mapping of a compact space X onto a Hausdorff space Y. Then Y has the identification topology determined by f.

As a further corollary, let $\pi_f:X \rightarrow X/\sim_f$ be the identification map which carries each element $x \in X$ into its equivalence set determined by the relation $x \sim_f x'$ if $f(x) = f(x')$. $\sim_f$ is continuous so $X/\sim_f$ is compact. By Theorem 8.2 of Chapter 3 there is a continuous map $f^*:X/\sim_f \rightarrow Y$ such that $f^*\pi_f = f$. As was remarked in that section, f^* is one-one; hence by Theorem 2.14, f^* is a homeomorphism.

Corollary 7.3 The mapping $f^*:X/\sim_f \rightarrow Y$ induced by a continuous function $f:X \rightarrow Y$ of a compact space onto a Hausdorff space is a homeomorphism.

One may think of a point $\bar{x} \in X/\sim_f$ as being represented by "pasting" together the various points in $\bar{x}$. As an example we shall consider a cylinder. We start with a rectangle with four corner vertices A, B, B', A' [see Figure 34a] and identify the edge AB with the edge $A'B'$ in such a way that A is identified with A' and B with B', then we obtain a surface that is homeomorphic to the cylinder in Figure 34b. We may equally well picture the cylinder as being the topological space obtained by replacing both A and A' by a new point A^*, both B and B' by a new point B^*, and similarly any pair of corresponding points C and C' on the respective edges AB and $A'B'$ is replaced by a new point C^* as indicated in Figure 34c.

Furthermore, a neighborhood of this new point C^* would contain the interior of the small semi-circles drawn in Figure 34c. It is interesting to note that if in this figure we join C^* to itself by the

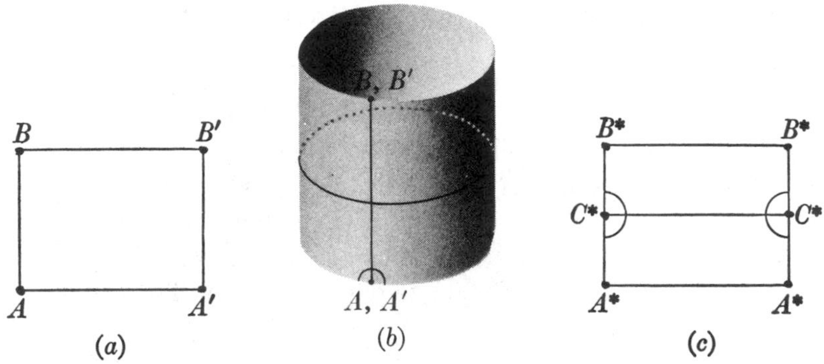

Figure 34

path represented by the horizontal line, the space consisting of the points of this line would be homeomorphic to a circle [such as the one drawn about the middle of the cylinder in Figure 34b], for it consists of an interval whose end points have been identified. This is a special case of the following general result.

LEMMA 7.4 Let X and Y be topological spaces, let $f:X \to Y$ be a continuous function that is onto, and let Y have the identification topology induced by f. If $B \subset Y$ is such that $A = f^{-1}(B)$ is closed, then the subspace B of Y has the identification topology induced by the restriction $f \mid A : A \to B$.

Proof. We must show that a subset F of B is closed in B if and only if $(f \mid A)^{-1}(F)$ is closed in A. The restriction $f \mid A$ of the continuous function f to $A = f^{-1}(B)$ is continuous, so that if F is closed in B, then $(f \mid A)^{-1}(F)$ is closed in A. Conversely, suppose that $(f \mid A)^{-1}(F)$ is closed in A. Then, since A is closed in X, $(f \mid A)^{-1}(F)$ is closed in X. If we prove that $(f \mid A)^{-1}(F) = f^{-1}(F)$, it will follow that F is closed in Y and consequently in B, for Y has the identification topology and therefore $f^{-1}(F)$ closed in X implies F closed in Y. It remains to prove $(f \mid A)^{-1}(F) = f^{-1}(F)$. Suppose that $x \in f^{-1}(F)$. To show that $x \in (f \mid A)^{-1}(F)$ we must show that $x \in A$ and $(f \mid A)(x) \in F$. But if $x \in f^{-1}(F)$, then $f(x) \in F \subset B$, whence $x \in f^{-1}(B) = A$. Thus x is in the domain of $f \mid A$ and $(f \mid A)(x) = f(x) \in F$, hence $x \in f^{-1}(F)$ implies that $x \in (f \mid A)^{-1}(F)$. Conversely, if $x \in (f \mid A)^{-1}(F)$, then $(f \mid A)(x) \in F$. Now $(f \mid A)(x) = f(x)$, thus $f(x) \in F$ and $x \in f^{-1}(F)$. It follows that $(f \mid A)^{-1}(F) = f^{-1}(F)$, and the proof is complete.

Another surface that may be obtained by identifying some of the boundary points of a rectangle is a surface called the Möbius strip or band. Starting again with the rectangle whose vertices we shall now label in the order A, B, A', B' [see Figure 35a], we identify the edge AB with the edge $B'A'$ by first giving the rectangular strip a 180 degree twist, so that the vertices A and A' coincide and the vertices B and B' coincide [Figure 35b]. A topologically equivalent space is indicated in Figure 35c, where corresponding or identified pairs of points such as A, A' have been replaced by a single new point A^*. The fact that we intend to identify the two edges AB and $A'B'$ of Figure 35a with a twist is often indicated by labelling the edges with the same letter, such as "a," and then placing arrowheads on these edges in such a position that the resulting identification

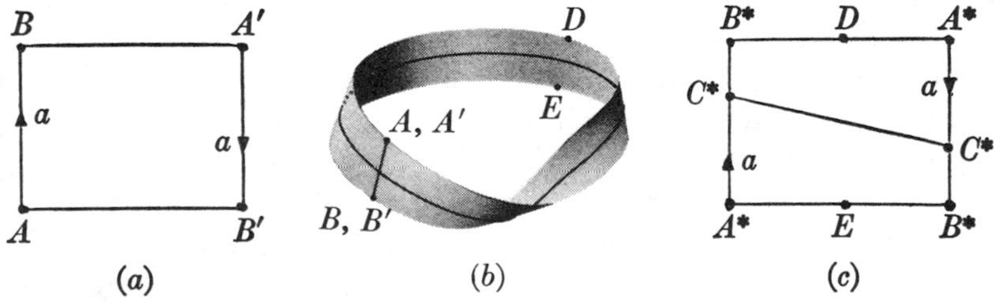

Figure 35

matches up or superimposes the two arrowheads. The Möbius strip has many curious properties. The oblique line in Figure 35c joining C^* to itself is homeomorphic to a circle. The upper horizontal line running from B^* through D to A^* is homeomorphic to an interval. However, if on the Möbius strip we trace out the curve from B^* through D to A^* and continue on [along the lower horizontal line of Figure 35c] through E back to B^* we trace out an interval with its end-points identified, that is, a circle. Thus the Möbius strip is a surface whose bounding curve is a circle. Other interesting properties may be deduced from the representation in Figure 35c. For example, if the Möbius strip is cut down its center, the resulting surface will not be disconnected for we may still connect a point of the upper half rectangle in Figure 35c to a point of the lower half rectangle by joining both of them to the bounding curve $B^*DA^*EB^*$.

If an arrowhead is placed on a circle we say that the circle is *oriented*. The sense of rotation indicated by the arrowhead is then called the *positive orientation* and the opposite sense of rotation is called the *negative orientation*. An oriented circle in the plane can be moved about in the plane in an arbitrary manner but will always be oriented in the same sense when it returns to its original position. For this reason the plane is said to be *orientable*. On the Möbius strip an oriented circle can be moved around the strip, say along the oblique line in Figure 35c with its center initially at C^*, and when it returns to its original position the orientation will have been reversed. Surfaces with this property are called *non-orientable*.

So far we have considered only surfaces resulting from the identification of a pair of edges of a rectangle. If we identify the edges of a rectangle according to the scheme indicated in Figure 36a, the resulting topological space is called a *torus*.

A torus is topologically the surface of a donut or a rubber tire, as indicated in Figure 36b. We may view the torus as being obtained in two steps. First, we identify the two opposite edges labelled a of the rectangle to obtain a cylinder, and second, we identify the two resulting circular edges (labelled b) of the cylinder to obtain the torus. The justification for breaking the identification up into two steps is contained in the following proposition.

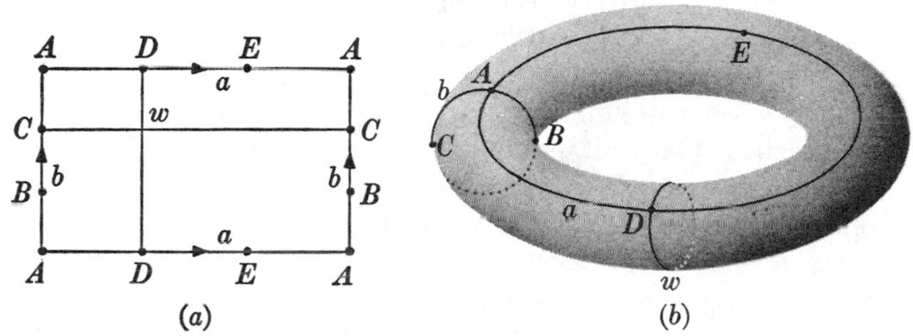

Figure 36

PROPOSITION 7.5 Let X, Y, Z be topological spaces, let $f: X \to Y$ and $g: Y \to Z$ be continuous and onto. If Y has the identification topology induced by $f: X \to Y$ and Z has the identification topology induced by $g: Y \to Z$, then Z has the identification topology induced by $gf: X \to Z$.

Proof. Clearly, if F is a closed subset of Z, then $(gf)^{-1}(F)$ is a closed subset of X, for gf is continuous. Conversely, suppose $(gf)^{-1}(F) = f^{-1}(g^{-1}(F))$ is a closed subset of X. Since Y has the identification topology induced by $f: X \to Y$, $g^{-1}(F)$ is a closed subset of Y. Similarly, since Z has the identification topology induced by $g: Y \to Z$, $g^{-1}(F)$ closed in Y implies that F is closed in Z. Thus F is closed if and only if $(gf)^{-1}(F)$ is closed; that is, Z has the identification topology induced by $gf: X \to Z$.

Topologically, the torus is the product of two circles. An arbitrary point w of the torus may be written as $w = (C, D)$, where C is a point of the circle b and D a point of the circle a in either Figure 36a or 36b. Furthermore, it is clear that the product of a neighborhood of C and a neighborhood of D is a neighborhood of w and conversely that a neighborhood of w contains the product of a neighborhood of C and a neighborhood of D. Thus the topology of the torus is the topology of the product of two circles. One would have anticipated this result if one viewed the torus as being generated by revolving a circle such as b in a circular path by moving it in such a way as to always have the point labelled A in contact with the circle labelled a.

There are two other surfaces resulting from the identification of opposite pairs of edges of a rectangle. One of these surfaces is called a *Klein bottle.* The Klein bottle may be obtained by first identifying the edges labelled a in Figure 37a in the prescribed manner to obtain a cylinder, and then identifying the two circles labelled b in either Figure 37a or 37b, not, however, in the manner

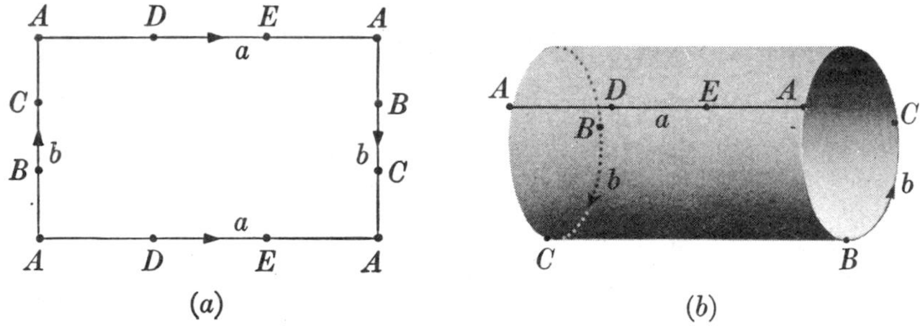

Figure 37

of Figure 36 to obtain a torus, but with a "twist." Unfortunately, at least from the point of view of our visualization of the Klein bottle, there is no way to identify these two circular edges of the cylinder of Figure 37b without forcing the surface of the Klein bottle to intersect or pass through itself. For this reason, it is helpful to construct the Klein bottle in several pieces.

In Figure 38 we have three rectangles. If the rectangles R_1 and R_2 are joined along the edge labelled c and the rectangles R_2 and R_3 are joined along the edge labelled d, we obtain the rectangle and identifications of Figure 37, so that Figure 38 also represents the Klein bottle. If, in these three rectangles, we first

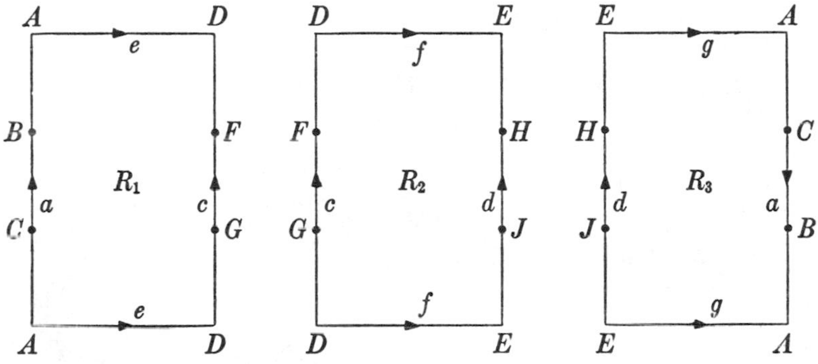

Figure 38

identify the pairs of edges labelled e, f, and g respectively, we obtain three cylinders that are homeomorphic to the three corresponding cylindrical surfaces of Figure 39, also labelled R_1, R_2, R_3. To construct the Klein bottle we need only identify these three cylinders along the pairs of circular edges labelled a, c, and d, respectively. We may join the cylinders R_1 and R_3 along the circles labelled a, so that R_3 lies inside R_1. If we then join R_1 and R_2 along

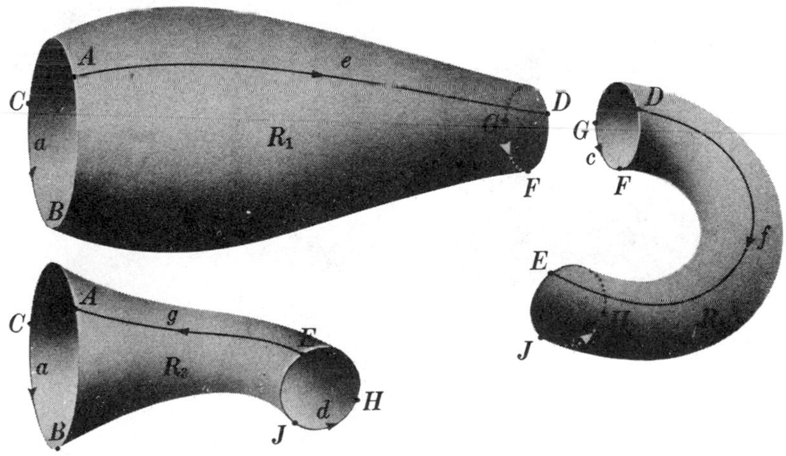

Figure 39

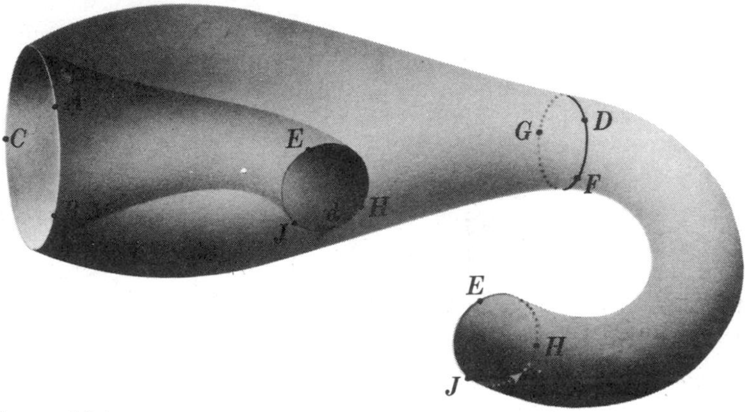

Figure 40

the circles labelled c, we obtain the cylinder pictured in Figure 40. To complete the construction, we must identify the two circles labelled d (Figure 40) in the prescribed manner. Any attempt to literally carry out this identification will be frustrated by our inability to pass through the surface of the cylinder. We must therefore either be content, as in Figure 40, to indicate this

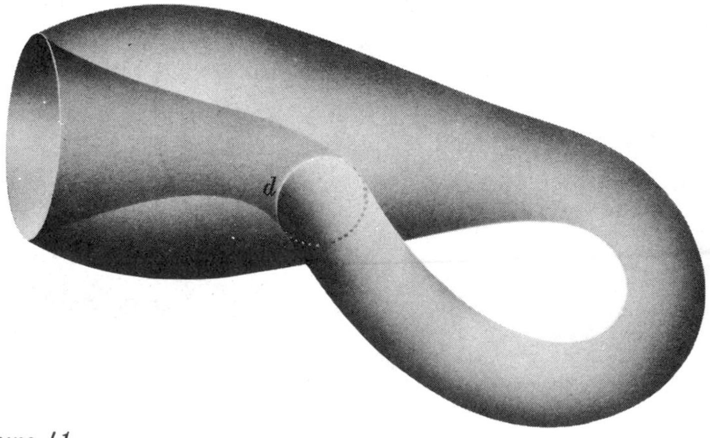

Figure 41

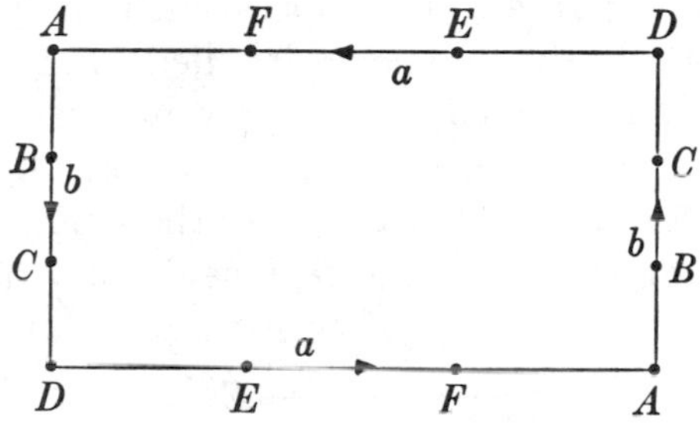

Figure 42

identification, or adopt the fiction that in Figure 41 the Klein bottle does not intersect itself along the circle d, but that each point along d is to represent at the same time two points of the Klein bottle.

The last surface we shall consider in detail is obtained by identifying both of the pairs of opposite edges of a rectangle with a "twist." These identifications are indicated in Figure 42. Note that in this figure all the vertices are not identified with one another, but only diagonally opposite vertices are joined together. In order to relate this surface to some of the preceding surfaces, we shall adopt the same method as the one used in the examination of the Klein bottle, [one might call this the "cut-and-paste method"]. We first separate the large rectangle into three smaller rectangles R_1, R_2, R_3, which when re-identified along the pairs of edges labelled c, d, will again give us the rectangle and the

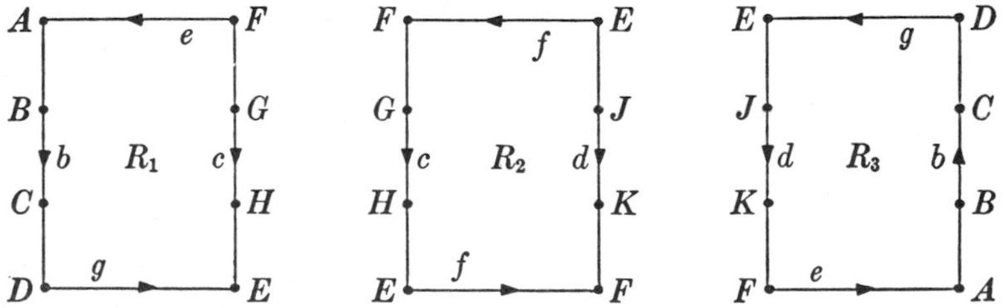

Figure 43

identifications of Figure 42. This operation is indicated in Figure 43. If we first join the two edges labelled f in rectangle R_2 we obtain a Möbius strip. Since we are only interested in the topological nature of this surface, we may distort [by homeomorphisms] the two rectangles R_1 and R_3 into the semicircular regions of Figure 44. If we then join the regions R_1 and R_3 along their

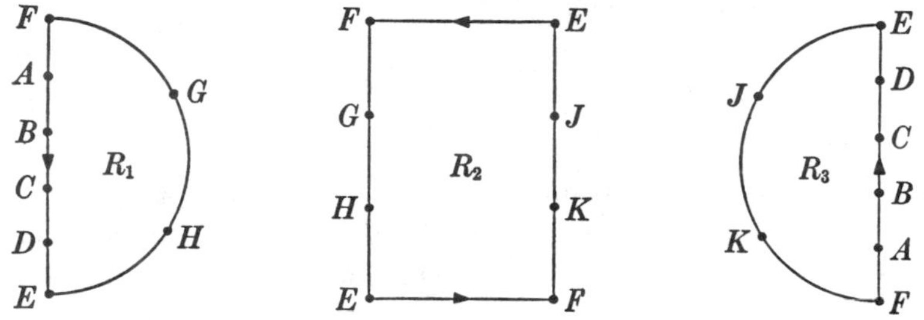

Figure 44

common edge $FABCDE$ we obtain the disc and the Möbius strip of Figure 45, with the indicated identifications. The surface we

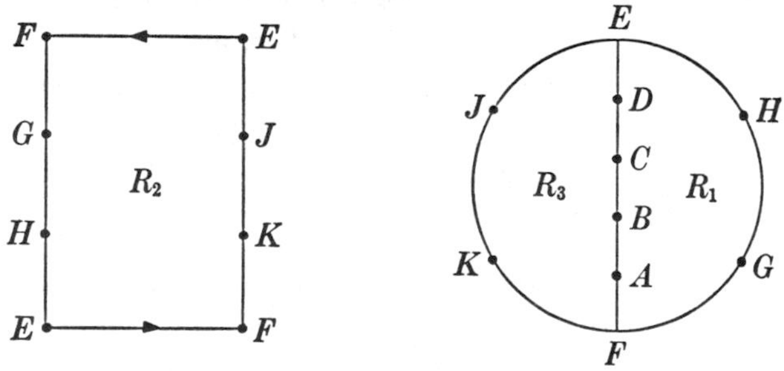

Figure 45

have been considering is therefore a Möbius strip whose boundary circle $FGHEJKF$ is to be attached to the boundary circle $FGHEJKF$ of a disc. This last surface is easily seen to be homeomorphic to one of the models of the "real projective plane," namely, a disc with antipodal points identified.

An analytic model of the real projective plane is obtained in the following manner. Let $A = R^3 - \{(0, 0, 0)\}$ be the set of all ordered triples (x_1, x_2, x_3) of real numbers such that not all of x_1, x_2, x_3 are zero. Define an equivalence relation on A by $(x_1, x_2, x_3) \sim (y_1, y_2, y_3)$, if there is a real number $r \neq 0$ such that $rx_1 = y_1$, $rx_2 = y_2$, $rx_3 = y_3$. The collection of equivalence sets P is the real projective plane. A point $p \in P$ is the collection of all points on a given straight line through the origin of R^3 other than the origin itself. The intersection of this equivalence set p with the unit sphere S^2 in R^3 is a pair of antipodal points. If we confine ourselves to the hemisphere of S^2 lying above the plane $x_3 = 0$, each equivalence set p meets the hemisphere in either a single point in the interior of the hemisphere or in a pair of antipodal points on the equator or boundary of the hemisphere. This upper hemisphere is a disc (view it from the point at the north pole so that it may be projected onto the equatorial plane). Identifying antipodal points on the boundary yields an identification space which is equivalent to the real projective plane.

The sphere, torus, Klein bottle, and projective plane are examples of a larger class of surfaces that may be obtained by identifying pairs of edges of a polygon with $2n$ sides. Such surfaces are called closed 2-manifolds. For example, in Figure 46 we have

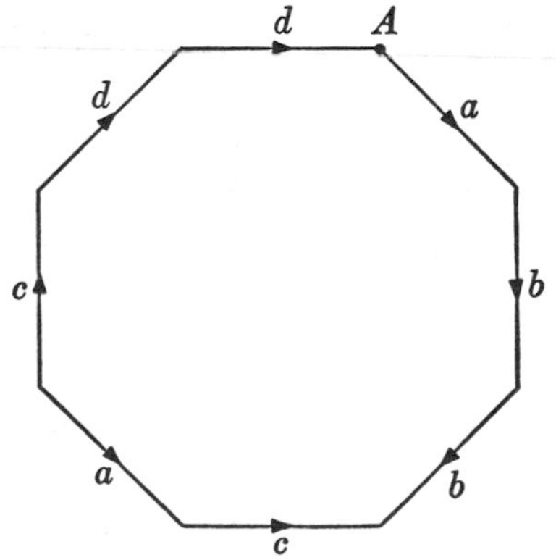

Figure 46

indicated a surface that can be obtained by identifying pairs of sides of an octagon. With each such figure we may associate a "surface symbol." We do so by starting at any vertex, such as A in Figure 46, and writing down the labels of the edges in clockwise order if the arrow along that edge is also pointing in the clockwise direction or the label with an inverse sign above if the arrow points in the counterclockwise direction. Thus a surface symbol for the surface of Figure 46 would be $abbc^{-1}a^{-1}cdd$. Referring back to Figure 36, one can see that a surface symbol for the torus is $ab^{-1}a^{-1}b$.

By the "cut-and-paste" method one can show that each 2-manifold is homeomorphic to a 2-manifold whose surface symbol is of one of the following four forms: $abb^{-1}a^{-1}$; $a_1b_1a_1^{-1}b_1^{-1} \ldots a_pb_pa_p^{-1}b_p^{-1}$, $p \geqq 1$; $abab$; $a_1a_1 \ldots a_qa_q$, $q > 1$. The first form indicates that the surface is homeomorphic to a sphere. The second form includes the surface symbol of a torus and in general indicates that the surface is homeomorphic to a sphere with p handles. These two classes of surface are orientable. They can all be constructed in three-dimensional Euclidean space. The third form indicates that the surface is homeomorphic to the projective plane. We have seen that the projective plane is a disc to whose circular boundary has been attached a Möbius strip. One may think of the disc as constituting the portion of the surface of a sphere obtained by removing a circular region. Attaching a Möbius strip to the circular boundary of this region is called attaching "a crosscap." Thus the projective plane is called "a sphere with crosscap." In the same manner, the fourth form consists of all surfaces obtained by attaching q Möbius strips or crosscaps to a sphere with q circular regions removed.

EXERCISES

1. Prove that the triangle T with two edges identified as in Figure 47 is homeomorphic to a disc.
2. Prove that the triangle S with two edges identified as in Figure 48 is a Möbius strip.

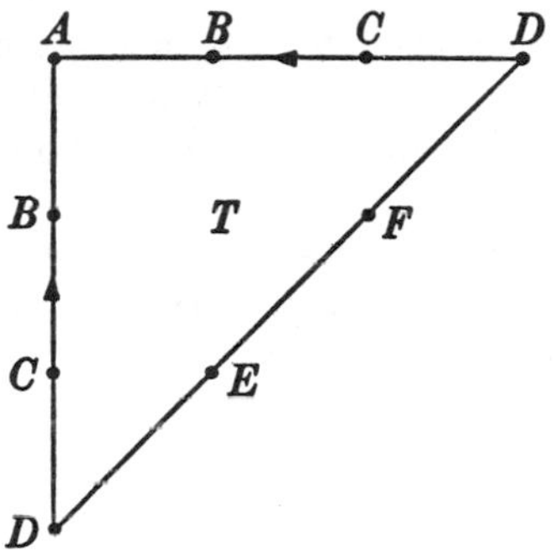

Figure 47

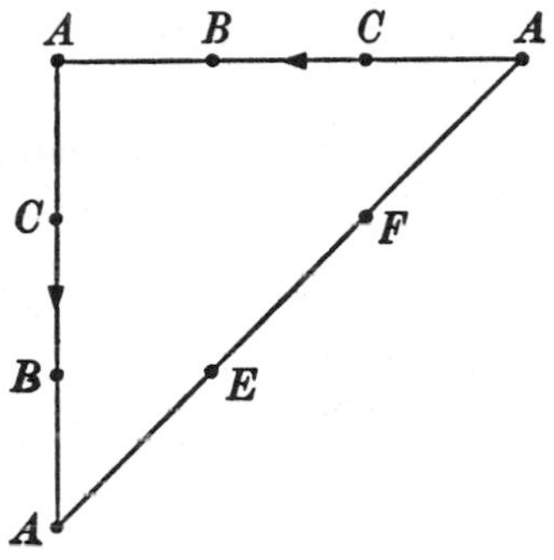

Figure 48

3. Prove that the Klein bottle is homeomorphic to a surface with surface symbol $a_1a_1a_2a_2$ by cutting the rectangle of Figure 49 along the diagonal c and pasting the resulting triangles along their common edge b.

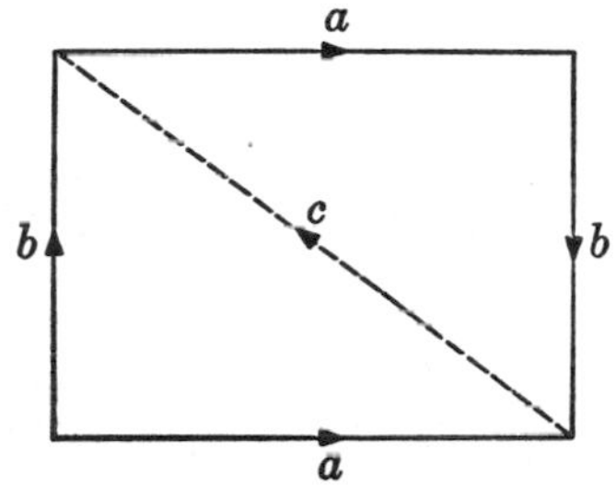

Figure 49

4. Show that if the Klein bottle of Figure 50 is cut along the curves c and d the result is two Möbius strips and that therefore the Klein bottle is two Möbius strips joined along their circular boundaries.

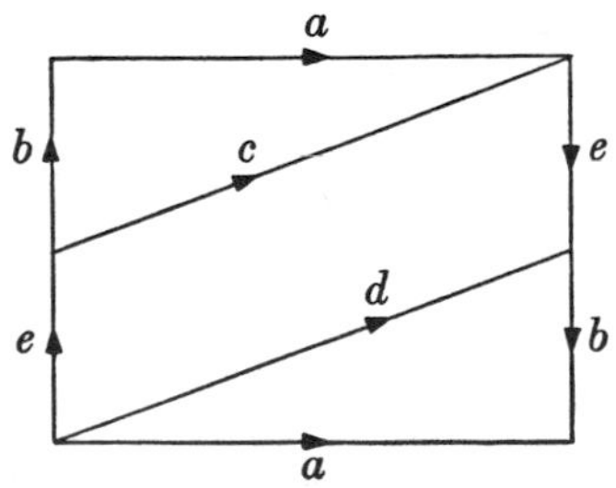

Figure 50

5. Cut the Klein bottle of Figure 51 along the curves c_1, c_2, c_3, c_4 and d_1, d_2. Show that the regions labelled S_1, S_2, S_3, S_4 are pasted together to form a surface homeomorphic to a cylinder and therefore homeomorphic to a sphere with two circular regions removed whose boundaries are the circles d_1d_2 and $c_1c_2c_3c_4$ respectively. Show that the region labelled M_1 is a Möbius strip whose boundary is d_1d_2 and that the regions labelled N_1 and N_2 form a second Möbius strip whose boundary is $c_1c_2c_3c_4$.

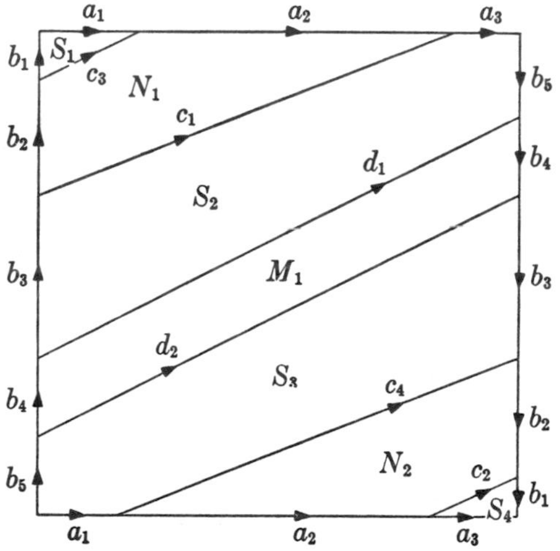

Figure 51

6. Prove that the following three statements about a closed 2-manifold are equivalent: (i) M contains a Möbius strip; (ii) M is non-orientable; (iii) the surface symbol of M contains two occurrences of some symbol "a."

For further reading in general topology we would include Kelley, Dugundji, or Simmons, while Blackett, *Elementary Topology,* Fréchet and Fan, *Combinatorial Topology,* Wallace, *Introduction to Algebraic Topology,* and Chinn and Steenrod, *First Concepts of Topology* are recommended for an introduction to topics in algebraic topology.

Bibliography

This list has been restricted to English texts and a few papers which would be appropriate to a student who has mastered the material in this book.

BING, R. H., "Elementary Point Set Topology." Herbert Ellsworth Slaught Memorial Papers, No. 8; *American Mathematical Monthly*, Vol. 67, No. 7, (1960).

BLACKETT, D. W., *Elementary Topology*. New York: Academic Press, 1967.

CHINN, W. G. and N. E. STEENROD, *First Concepts of Topology*. New York: Random House, 1966.

COXETER, H. M. S., "Map-Coloring Problems." *Scripta Mathematica*, Vol. XXIII (1957), pp. 11–25.

CROWELL, R. H. and R. H. FOX, *Introduction to Knot Theory*. New York: Blaisdell, 1963.

DUGUNDJI, J., *Topology*. Boston, Mass.: Allyn and Bacon, 1965.

FRÉCHET, M. and K. FAN, *Combinatorial Topology*. Boston, Mass.: Prindle, Weber and Schmidt, 1967.

GREENBERG, M., *Lectures on Algebraic Topology*. New York: W. A. Benjamin, Inc., 1967.

GREEVER, J., *Theory and Examples of Point-Set Topology*. Belmont, Cal.: Brooks/Cole, 1967.

HALMOS, P. R., *Naive Set Theory*. New York: Van Nostrand Reinhold, 1960.

HILTON, P. J., *An Introduction to Homotopy Theory*. Cambridge: Cambridge University Press, 1953.

HOCKING, J. G. and G. S. YOUNG, *Topology*. Reading, Mass.: Addison-Wesley, 1961.

HU, S. T., *Elements of General Topology*. San Francisco, Cal.: Holden-Day, 1964.

HU, S. T., *Introduction to General Topology*. San Francisco, Cal.: Holden-Day, 1966.

HUREWICZ, W. and H. WALLMAN, *Dimension Theory*. Princeton, N. J.: Princeton University Press, 1941.

KAPLANSKY, I., *Set Theory and Metric Spaces*. Boston: Allyn and Bacon, 1972.

KELLEY, J. L., *General Topology*. New York: Van Nostrand, 1955.

KOLMOGOROV, A. N. and S. V. FOMIN, *Elements of the Theory of Functions and Functional Analysis*. Baltimore: Graylock Press, 1957 (vol. 1); 1961 (vol. 2).

KURATOWSKI, K., *Introduction to Set Theory and Topology*. Reading, Mass.: Addison-Wesley, 2nd ed., 1972.

LEFSHETZ, S., *Introduction to Topology*. Princeton, N. J.: Princeton University Press, 1949.

MASSEY, W. S., *Algebraic Topology, An Introduction*. New York: Harcourt, Brace & World, Inc., 1967.

McCARTY, G., *Topology*. New York: McGraw-Hill, 1967.

ORE, O., *Graphs and Their Uses*. New York: Random House, 1963.

PERVIN, W. J., *Foundations of General Topology*. New York: Academic Press, 1964.

SEEBACH, J. and L. A. STEEN, *Counterexamples in Topology*. New York: Holt, Rinehart and Winston, Inc., 1970.

SIMMONS, G. F., *Introduction to Topology and Modern Analysis*. New York: McGraw-Hill, 1963.

SINGER, I. M. and J. A. THORPE, *Lecture Notes on Elementary Topology and Geometry*. Glenview, Ill.: Scott, Foresman and Company, 1967.

TUCKER, A. W., "Some Topological Properties of Disk and Sphere," *Proceedings of the First Canadian Mathematical Congress*, pp. 285–309, Montreal, Canada, 1945.

WALL, C. T. C., *A Geometric Introduction to Topology*. Reading, Mass.: Addison-Wesley, 1972.

WALLACE, A. H., *Introduction to Algebraic Topology*. New York: Pergamon, 1957.

WILLARD, S., *General Topology*. Reading, Mass.: Addison-Wesley Publishing, 1970.

Index

Accumulation point, 172
Annulus, 139
Antipodal point, 127
Archimedean principle, 87
Automorphism, 155

Banach space, 179
Basic neighborhood space, 81
Basis for neighborhood system, 45, 80, 99
Basis for open sets, 57, 98
Bolzano-Weierstrass property, 175–176, 183–184
Borsuk-Ulam theorem, 128
Bound
  greatest lower, 49
  least upper, 49
  lower, 49
  upper, 49
Boundary, 85
Brouwer fixed-point theorem, 125

Cantor set, 101
Cartesian product, 9
Category, 107
Cauchy sequence, 178
Cauchy's inequality, 66
Characteristic function, 15
Closure, 82
Compact, 159
  locally, 171
Compactum, 172
Complement, 5
Completely separable, 184
Completeness postulate, 49
Complete system of neighborhoods, 41, 76
Component, 131
  of a point, 130

Cone, 106
Connected, 113, 121
  locally, 132
  path-, 134
Contractive, 186
Countable, 180
Covering, 158
  finite, 158
  open, 159
Crosscap, 198
Cylinder, 187

DeMorgan's laws, 5, 8
Dense, 178
Denumerable, 180
Diagram, 19
  commutative, 20
Direct product, 10
Disc, 126
Disconnected, 113
Distance
  between a point and a set, 50
  Euclidean, 33
  function, 30

Equivalence class, 16
Euclidean n-space, 11, 72

Finite intersection property, 164
First axiom of countability, 185
Fixed point, 123–127

Function
  bijective, 13
  bounded, 34
  composition of, 17, 18
  constant, 14
  continuous, 36, 37, 88
  continuous at a point, 36, 88
  domain of a, 13
  extension of a, 23
  identity, 14
  inclusion, 24
  injective, 13
  inverse, 21, 22
  invertible, 21
  one-one, 13, 22
  onto, 13, 22
  range of a, 13
  restriction of a, 23
  successor, 1
  surjective, 13
Functor, 107

Graph, 12
Group, 108

Hausdorff
  space, 76
  axiom, 76
Heine-Borel theorem, 167
Hilbert space, 66
Homeomorphism, 90
Homomorphism, 108
Homotopic
  paths, 141
  functions, 150

Homotopy, 141
  class of paths, 145

Identification, 101
Image, 13
Inclusion mapping, 24
Indexed family, 7
  product of an, 26
Interior, 84
Intermediate value theorem,
    122
Intersection, 4, 7
Interval, 119, 121
  closed, 3
  open, 3
Inverse image, 13
Isolated point, 55
Isometric, 60
Isomorphic, 150, 155
Isomorphism, 155

Klein bottle, 192

Lebesgue number, 175
  limit
    of a sequence, 47
    point, 54
Lindelöf's theorem, 184
Loop, 135

Mapping, 13
  open, 89
(*see also* Function)

Metrically equivalent, 60
Metric space, 30
  complete, 178
Möbius strip, 189

Natural numbers, 1–2
Neighborhood, 41, 73
  space, 77
  relative, 93
Non-orientable, 190
n-sphere, 59, 127
n-tuple, 11

One-point compactification,
    172
Open ball, 40
Ordered pair, 9
Orientable, 190
Oriented, 190

Partition, 131
Path, 134
  closed, 135
  component, 138
  constant, 145
  initial point of a, 134
  product of, 145
  terminal point of a, 134
Path-connected, 134
Peano's axioms, 1
Positive integers, 1
Product of sets, 26
Product of spaces, 98

Projection, 16, 26
  first, 15
  second, 15
Projective plane, 195–197

Quotient, 16

Rational density theorem, 87
Real line, 72
Refinement, 185
Relation, 15
  equivalence, 16
  reflexive, 16
  symmetric, 16
  transitive, 16
Relatively closed, 95
Relatively open, 92
Relative neighborhood, 93
Relative topology, 93
Retract, 150
  deformation, 150

Schwarz's lemma, 66
Second axiom of countability, 184
Separable, 185
Set, 2
  bounded, 49, 164
  closed, 54, 74
  empty, 3
  null, 3
  open, 52, 71, 77
Simply connected, 151
Subcovering, 158
Subset, 3
  improper, 3
  proper, 3
Subspace, 58, 92
Surface symbol, 198
Suspension, 106

Topologically equivalent, 62
Topological property, 115
Topological space, 71
  associated with a metric space, 71
Topology, 71
  discrete, 72
  identification, 102
Torus, 106, 190
Transformation, 13

Uncountable, 181
Uniformly continuous, 177
Unit n-cube, 59, 171
Union
  of two sets, 4, 7
  of an indexed family of sets, 7

Venn diagrams, 5

Weaker, 95

A CATALOG OF SELECTED

DOVER BOOKS

IN SCIENCE AND MATHEMATICS

Math–Geometry and Topology

ELEMENTARY CONCEPTS OF TOPOLOGY, Paul Alexandroff. Elegant, intuitive approach to topology from set-theoretic topology to Betti groups; how concepts of topology are useful in math and physics. 25 figures. 57pp. 5⅜ x 8½. 0-486-60747-X

COMBINATORIAL TOPOLOGY, P. S. Alexandrov. Clearly written, well-organized, three-part text begins by dealing with certain classic problems without using the formal techniques of homology theory and advances to the central concept, the Betti groups. Numerous detailed examples. 654pp. 5⅜ x 8½. 0-486-40179-0

EXPERIMENTS IN TOPOLOGY, Stephen Barr. Classic, lively explanation of one of the byways of mathematics. Klein bottles, Moebius strips, projective planes, map coloring, problem of the Koenigsberg bridges, much more, described with clarity and wit. 43 figures. 210pp. 5⅜ x 8½. 0-486-25933-1

THE GEOMETRY OF RENÉ DESCARTES, René Descartes. The great work founded analytical geometry. Original French text, Descartes's own diagrams, together with definitive Smith-Latham translation. 244pp. 5⅜ x 8½. 0-486-60068-8

EUCLIDEAN GEOMETRY AND TRANSFORMATIONS, Clayton W. Dodge. This introduction to Euclidean geometry emphasizes transformations, particularly isometries and similarities. Suitable for undergraduate courses, it includes numerous examples, many with detailed answers. 1972 ed. viii+296pp. 6⅛ x 9¼. 0-486-43476-1

PRACTICAL CONIC SECTIONS: THE GEOMETRIC PROPERTIES OF ELLIPSES, PARABOLAS AND HYPERBOLAS, J. W. Downs. This text shows how to create ellipses, parabolas, and hyperbolas. It also presents historical background on their ancient origins and describes the reflective properties and roles of curves in design applications. 1993 ed. 98 figures. xii+100pp. 6½ x 9¼. 0-486-42876-1

THE THIRTEEN BOOKS OF EUCLID'S ELEMENTS, translated with introduction and commentary by Sir Thomas L. Heath. Definitive edition. Textual and linguistic notes, mathematical analysis. 2,500 years of critical commentary. Unabridged. 1,414pp. 5⅜ x 8½. Three-vol. set.
Vol. I: 0-486-60088-2 Vol. II: 0-486-60089-0 Vol. III: 0-486-60090-4

SPACE AND GEOMETRY: IN THE LIGHT OF PHYSIOLOGICAL, PSYCHOLOGICAL AND PHYSICAL INQUIRY, Ernst Mach. Three essays by an eminent philosopher and scientist explore the nature, origin, and development of our concepts of space, with a distinctness and precision suitable for undergraduate students and other readers. 1906 ed. vi+148pp. 5⅜ x 8½. 0-486-43909-7

GEOMETRY OF COMPLEX NUMBERS, Hans Schwerdtfeger. Illuminating, widely praised book on analytic geometry of circles, the Moebius transformation, and two-dimensional non-Euclidean geometries. 200pp. 5⅝ x 8¼. 0-486-63830-8

DIFFERENTIAL GEOMETRY, Heinrich W. Guggenheimer. Local differential geometry as an application of advanced calculus and linear algebra. Curvature, transformation groups, surfaces, more. Exercises. 62 figures. 378pp. 5⅜ x 8½. 0-486-63433-7

Mathematics

FUNCTIONAL ANALYSIS (Second Corrected Edition), George Bachman and Lawrence Narici. Excellent treatment of subject geared toward students with background in linear algebra, advanced calculus, physics and engineering. Text covers introduction to inner-product spaces, normed, metric spaces, and topological spaces; complete orthonormal sets, the Hahn-Banach Theorem and its consequences, and many other related subjects. 1966 ed. 544pp. 6⅛ x 9¼. 0-486-40251-7

ASYMPTOTIC EXPANSIONS OF INTEGRALS, Norman Bleistein & Richard A. Handelsman. Best introduction to important field with applications in a variety of scientific disciplines. New preface. Problems. Diagrams. Tables. Bibliography. Index. 448pp. 5⅜ x 8½. 0-486-65082-0

VECTOR AND TENSOR ANALYSIS WITH APPLICATIONS, A. I. Borisenko and I. E. Tarapov. Concise introduction. Worked-out problems, solutions, exercises. 257pp. 5⅜ x 8¼. 0-486-63833-2

AN INTRODUCTION TO ORDINARY DIFFERENTIAL EQUATIONS, Earl A. Coddington. A thorough and systematic first course in elementary differential equations for undergraduates in mathematics and science, with many exercises and problems (with answers). Index. 304pp. 5⅜ x 8½. 0-486-65942-9

FOURIER SERIES AND ORTHOGONAL FUNCTIONS, Harry F. Davis. An incisive text combining theory and practical example to introduce Fourier series, orthogonal functions and applications of the Fourier method to boundary-value problems. 570 exercises. Answers and notes. 416pp. 5⅜ x 8½. 0-486-65973-9

COMPUTABILITY AND UNSOLVABILITY, Martin Davis. Classic graduate-level introduction to theory of computability, usually referred to as theory of recurrent functions. New preface and appendix. 288pp. 5⅜ x 8½. 0-486-61471-9

ASYMPTOTIC METHODS IN ANALYSIS, N. G. de Bruijn. An inexpensive, comprehensive guide to asymptotic methods–the pioneering work that teaches by explaining worked examples in detail. Index. 224pp. 5⅜ x 8½ 0-486-64221-6

APPLIED COMPLEX VARIABLES, John W. Dettman. Step-by-step coverage of fundamentals of analytic function theory–plus lucid exposition of five important applications: Potential Theory; Ordinary Differential Equations; Fourier Transforms; Laplace Transforms; Asymptotic Expansions. 66 figures. Exercises at chapter ends. 512pp. 5⅜ x 8½. 0-486-64670-X

INTRODUCTION TO LINEAR ALGEBRA AND DIFFERENTIAL EQUATIONS, John W. Dettman. Excellent text covers complex numbers, determinants, orthonormal bases, Laplace transforms, much more. Exercises with solutions. Undergraduate level. 416pp. 5⅜ x 8½. 0-486-65191-6

RIEMANN'S ZETA FUNCTION, H. M. Edwards. Superb, high-level study of landmark 1859 publication entitled "On the Number of Primes Less Than a Given Magnitude" traces developments in mathematical theory that it inspired. xiv+315pp. 5⅜ x 8½. 0-486-41740-9